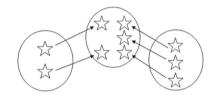

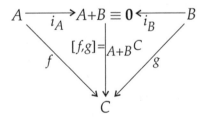

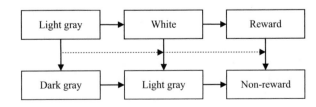

Mathematics in Mind

Series Editor
Marcel Danesi, *University of Toronto, Canada*

Editorial Board
Louis Kauffman, *University of Illinois at Chicago, USA*
Dragana Martinovic, *University of Windsor, Canada*
Yair Neuman, *Ben-Gurion University of the Negev, Israel*
Rafael Núñez, *University of California, San Diego, USA*
Anna Sfard, *University of Haifa, Israel*
David Tall, *University of Warwick, United Kingdom*
Kumiko Tanaka-Ishii, *Kyushu University, Japan*
Shlomo Vinner, *Hebrew University, Israel*

The monographs and occasional textbooks published in this series tap directly into the kinds of themes, research findings, and general professional activities of the Fields Cognitive Science Network, which brings together mathematicians, philosophers, and cognitive scientists to explore the question of the nature of mathematics and how it is learned from various interdisciplinary angles.

This series covers the following complementary themes and conceptualizations:

· Connections between mathematical modeling and artificial intelligence research; math cognition and symbolism, annotation, and other semiotic processes; and mathematical discovery and cultural processes, including technological systems that guide the thrust of cognitive and social evolution

· Mathematics, cognition, and computer science, focusing on the nature of logic and rules in artificial and mental systems

· The historical context of any topic that involves how mathematical thinking emerged, focusing on archeological and philological evidence

· Other thematic areas that have implications for the study of math and mind, including ideas from disciplines such as philosophy and linguistics

The question of the nature of mathematics is actually an empirical question that can best be investigated with various disciplinary tools, involving diverse types of hypotheses, testing procedures, and derived theoretical interpretations. This series aims to address questions of mathematics as a unique type of human conceptual system versus sharing neural systems with other faculties, whether it is a series-specific trait or exists in some other form in other species, what structures (if any) are shared by mathematics language, and more.

Data and new results related to such questions are being collected and published in various peer-reviewed academic journals. Among other things, data and results have profound implications for the teaching and learning of mathematics. The objective is based on the premise that mathematics, like language, is inherently interpretive and explorative at once. In this sense, the inherent goal is a hermeneutical one, attempting to explore and understand a phenomenon—mathematics—from as many scientific and humanistic angles as possible.

More information about this series at http://www.springer.com/series/15543

Yair Neuman

Mathematical Structures
of Natural Intelligence

 Springer

Yair Neuman
The Department of Brain and Cognitive Sciences
and the Zlotowski Center for Neuroscience
Ben-Gurion University of the Negev
Beer-Sheva, Israel

ISSN 2522-5405   ISSN 2522-5413   (electronic)
Mathematics in Mind
ISBN 978-3-319-88570-4   ISBN 978-3-319-68246-4   (eBook)
https://doi.org/10.1007/978-3-319-68246-4

Printed on acid-free paper

This Springer imprint is published by Springer Nature
The registered company is Springer International Publishing AG
The registered company address is: Gewerbestrasse 11, 6330 Cham, Switzerland

*This book is dedicated to my dear parents.
To my father, Zvi Neuman, who taught me a
love of serious books, a passion for truth and
that it is not only salmon that should swim
against the stream
and
to my mother, Yocheved Neuman, who taught
me that books can be fun, that truth
sometimes resides among other human
beings, and that a salmon might end its life
on a plate.*

Preface

For many years, I have been interested in identifying general structures underlying the various expressions of the human mind, from psychology to art and language. This interest has naturally led me to an intensive reading of texts such as those of the gestalt psychologists and of Jean Piaget and Gregory Bateson, whose interdisciplinary approach has had a great influence on my work. The idea of writing a book on "mathematical structures" of "natural intelligence" has slowly grown in my mind, always accompanied by two main concerns.

The first concern is that, in the postmodern and post-structuralist phase of the Western "intelligentsia," such a venture might be considered both anachronistic and pretentious. I have tried to address these concerns first by presenting a *neo*-structuralist approach that addresses some of the difficulties that were associated with the old structuralist venture and second by putting aside unjustified pretensions toward developing a grand theory. In contrast with some such grand theories of the past, this book is framed from a much more *reflectively* humble perspective.

The second major concern that bothered me is that, in the context of studying the human mind, "mathematical structures" may be a nice metaphor in some cases but it can also be an empty abstraction. Learning about past attempts to "mathematicize" the human mind, such as those of Kurt Lewin and his field theory, has left me with a strong and sour taste of disappointment. Mathematics is such an abstract field that, beyond the powerful *technical* tools it provides for outsiders, its ability to provide informative *structures* for modeling the human mind is extremely limited. We may enjoy ideas such as "topological psychology," but these intellectual games seem to lead nowhere.

In this book, I use the term "mathematical structures" to describe in an abstract language several configurations that may be highly informative in illuminating some aspects of our mind in fields ranging from neuroscience to poetry. The reader may therefore not find herein any attempt to introduce *the* "mathematics of natural intelligence." However, I still aim to identify general abstract structures that may be used for modeling "natural intelligence" (a term that is briefly introduced below and then clarified in Chap. 1).

Having said that, and being highly cognizant of the limits of a neo-structuralist venture, the final decision to write the book crystallized in my mind when I visited the beautiful city of Naples for a conference. One night, and while sitting near the window of my hotel room and observing the sea with its repeating patterns of waves, some threads wove together in my mind. I suddenly realized how deeply interconnected are several ideas that have interested me for many years, from Russell's definition of a number to the mathematical concept of the groupoid. From there, this book naturally emerged in the form you are currently reading.

Given this context, the book's organization is as follows. The first part of the book introduces the justification for studying general structures that we may use to model "natural intelligence." The term "intelligence" is used in the broad sense of *computing patterns*, and the adjective "natural" is used to draw a boundary between intelligence as it is evident in natural living systems and intelligence as it is studied in artificial systems and models. In fact, computation is the leading metaphor for studying the mind (Crane, 2015), and the inevitable question is: What is the difference between mathematical and computational modeling of the mind? This is quite a serious question that deserves an in-depth discussion which is beyond the scope of this book. However, while computational models of the mind attempt to understand human cognition in metaphorical/analogical terms of an abstract Turing machine, and/or to produce computational models of specific cognitive processes, in this book, I model human cognition by drawing on the very abstract theorization of category theory. Differences and similarities between computational models and the mathematical models presented in this book can be easily found, but the important point is that my theoretical point of departure for understanding the human mind is totally different.

I present Piaget's heroic venture to lay the grounds for a theory of structure but critically point to his failure. To address Piaget's failure, I introduce and use the language of category theory, a field of mathematics that has great relevance as a powerful tool for building models. The book is self-contained, in that it doesn't assume any prior knowledge of mathematics in general or of category theory in particular. However, I am aware of the fact that the book is extremely challenging in its level of theorization and abstraction and that it will require effort from the reader to struggle with some of its abstract formulations. I believe, though, that the intellectual benefits of these efforts are justified. Using category theory, I argue that a certain mathematical structure – the groupoid – may be used to address Piaget's main problem and to serve as a building block of structure. To illustrate the explanatory power of this conceptualization, I use the field of neuroscience and explore an unresolved question, which is why we observe local and recurrent cortical circuits in the mammalian brain.

The second part of the book aims to take us a step further by delving deeper into the power of category theory's conceptualization in modeling structure. I explain why natural intelligence is deeply "relational" and how structures such as *natural transformation* may model these relational processes. Moreover, I explain that natural intelligence is not only relational but also value laden and explore the gestalt aspect of structure by using a variety of category theory tools. In this part, I also introduce a novel principle that I describe as the natural transformation modeling principle and explain how it can serve as an alternative concept for studying uncertainty "in the wild."

The third and concluding part of the book is the least mathematical and involves the application of the general principles and ideas presented in the first two parts, in order to reach a better understanding of the three challenging aspects of the human mind. In this part of the book, I try to explain the human representation of the number system and specifically why our counting ability is different from the ability evident among non-human organisms and why it is so difficult to grasp the idea of zero. I also model the process of analogical reasoning and metaphor by pointing to its underlying relational structure and its deep grounding in episodic memory and in cultural semiotic threads necessary for understanding the complexity of metaphors. Taking this idea a step forward, in this third section of the book and following the theory of Ignacio Matte-Blanco, I adopt the idea of the "unconscious" as expressing creative processes of symmetrization and illustrate how deeply connected these processes are to the general themes presented in the book and how they may be used to broaden our understanding of metaphorical processes and the creativity of the human mind.

As the reader may have already realized, the book is intellectually challenging, and for this reason, only the efforts required to cross the bridge of mathematical abstraction seem to be fully justified. To ease the cognitive load the reader may experience, I have added a bullet summary at the end of each chapter. In addition, I have not overloaded the reader with references or footnotes, which are sometimes used to express the author's "deep mastery" in his academic field.

An etymology[1] of "mastery" in the English language explains that mastery appeared as a transitive verb (i.e., a verb with a defined object) in 1225 and was used in the sense of "overcoming," in the military sense of defeating an enemy. The learned scholar, the "úþwita" (Old English), who is beyond (i.e., úþ) the ordinary wisdom of human beings is therefore the one who "defeats" and "conquers" the object of knowledge. But can we really conquer knowledge?

The old Jewish rabbis who read the book of Ecclesiastes encountered the following text: "All the rivers run into the sea, yet the sea is not full; unto the place whither the rivers go, thither they go again." They metaphorically interpreted "rivers" as "wisdom" and "sea" as corresponding to one's heart. According to this interpretation, one may gain wisdom, but wisdom cannot be conquered, defeated, or "mastered" as if it were a beast to be tamed. Throwing the metaphor of mastery aside, we may step into the river.

Beer-Sheva, Israel   Yair Neuman

---

[1] All references to the etymology of words are drawn from *The Historical Thesaurus of English* unless stated otherwise.

Acknowledgments

A couple of years ago, a psychologist (Jaan Valsiner) and a mathematician (Lee Rudolph) convinced me to publish a paper on category theory and the mind, despite my concerns that such an abstract paper might be like a tree falling in a wood with no ears to hear it. The encouragement of Jaan and Lee should be acknowledged and praised as a model of what the scientific milieu should look like. In addition, a sabbatical year at the Weizmann Institute of Science gave me the opportunity to dedicate intensive time to learning and thinking. My conversations with my host, Rafi Malach, have enormously contributed to the emergence of this book. Rafi's position, both knowledgeable and skeptical, contributed to the chapter that deals with neural networks (Chap. 5), which was originally designed as a coauthored paper. This chapter couldn't have been written without him, and his intellectual challenges have put their mark on this book. Finally, Marcel Danesi, who invited me to contribute this book to the series of which it is now a part, didn't leave me with any option other than a positive one. I first met Marcel at 2000, and since then, I have been impressed by his energy, his academic enterprises, and his encouraging and always supportive approach toward younger researchers and academic innovations. Finally, I would like to thank my editor – Hazel Bird – for professionally editing this book according to the highest standards and my Springer editor Elizabeth Loew for her warm support of this project.

Contents

Part I

1 Introduction: The Highest Faculty of the Mind   3
   Summary   11

2 What Is Structure? Piaget's Tour de Force   13
   Piaget on Structure   14
   Reversibility and Irreversibility   25
   Summary   29

3 Category Theory: Toward a Relational Epistemology . .   31
   Universal Constructions   36
   The Co-product   41
   Summary   46

4 How to Trick the Demon of Entropy   47
   Summary   51

5 Neural Networks and Groupoids   53
   Summary   62

Part II

6 Natural Intelligence in the Wild   65
   Summary   70

7 Natural Intelligence Is About Meaning   71
   Summary   76

8   From Identity to Equivalence   77
      Summary   84

9   On Negation   85
      Summary   91

10   Modeling: The Structuralist Way   93
      Summary   102

11   On Structures and Wholes   103
      Summary   117

Part III

12   Let's Talk About Nothing: Numbers and Their Origin .   121
      Summary   130

13   King Richard Is a Lion: On Metaphors and Analogies .   131
      Summary   146

**14   The Madman and the Dentist: The Unconscious
      Revealed**   147
      Summary   154

15   Discussion   155

References   163

About the Author   167

Author Index   169

Subject Index   171

List of Figures

Fig. 2.1 Smiley's face 16
Fig. 2.2 Smiley's disorganized face 16

Fig. 3.1 A map illustrating the permutation of three boxes 32
Fig. 3.2 A structure with three isomorphic objects 38
Fig. 3.3 Projections of A × B 39
Fig. 3.4 The commuting diagram 39
Fig. 3.5 The taxonomy of Chair 40
Fig. 3.6 The product as a pattern 41
Fig. 3.7 The co-product 41
Fig. 3.8 The Würstchen and the Hotdog 45

Fig. 4.1 The product's projections 48
Fig. 4.2 Product and identity 48
Fig. 4.3 The co-product of sets 48
Fig. 4.4 The proof 48
Fig. 4.5 The groupoid 49
Fig. 4.6 A two-cell network 49

Fig. 6.1 An apple in transformation 67
Fig. 6.2 The first three phases of consumption 67
Fig. 6.3 The consumption as a groupoid 67
Fig. 6.4 A functor between the transformations functors 68

Fig. 7.1 The mapping between the differences and value
categories 74

Fig. 8.1 Functors between the apples.. 79
Fig. 8.2 A functor between functors F and G 80
Fig. 8.3 Natural transformations... 81
Fig. 8.4 Two mappings from the first appearance
             of the apple 82
Fig. 8.5 Adjointness.. 84
Fig. 8.6 Adjointness in Köhler's hen experiment.................... 84

Fig. 9.1 The sub-object diagram 88

Fig. 11.1 A dragonfly.. 105
Fig. 11.2 Sub-objects of the dragonfly..................................... 106
Fig. 11.3 The center of the dragonfly....................................... 107
Fig. 11.4 A checkerboard.. 108
Fig. 11.5 A rotating disk 111
Fig. 11.6 A diagram of US elections... 112
Fig. 11.7 The eye 117

Fig. 12.1 The formation of a number.. 125

Fig. 13.1 The structure of analogical reasoning........................ 134
Fig. 13.2 The formation of a metaphor 144

Fig. 14.1 A triangle 152

List of Table

Table 3.1 The context vectors of Sausage and Hotdog 43

Part I

Chapter 1
Introduction: The Highest Faculty of the Mind

Today, the term "intelligence" is usually discussed in its most simple sense as human "wisdom," which is measured through intelligence tests. However, the term "intelligence" dates back to the fourteenth century, when it was used to signify "the highest faculty of the mind, capacity for comprehending *general truths*" (*Online Etymological Dictionary*, emphasis mine).

One may hardly find evidence for the capacity for comprehending "general truths" in modern intelligence tests nor for the contention that this faculty is unique to human beings. In fact, the more we learn about the behavior of other organisms and living systems, from slime mold to the immune system, the more humble and open-minded we should be in discussing human intelligence and its uniqueness.

The *Historical Thesaurus of English* teaches us that the term "intelligence" has multiple origins that include various senses such as the communication of information and the characteristics of spiritual beings. The impression formed from perusing these senses is that the capacity to gain general "truths" has perceptual (i.e., gaining knowledge through the senses), social (i.e., gaining knowledge between people), and "divine" dimensions.

The association of intelligence with the divine may be clear even to the (post-) modern secular person; when witnessing the acquisition or the revelation of "general truths," whether in others, through observation, or in ourselves through introspection, we cannot avoid a deep sense of awe – probably the same sense evoked in our ancestors by the association between intelligent activity and the divine. This feeling is

© Springer International Publishing AG 2017
Y. Neuman, *Mathematical Structures of Natural Intelligence*, Mathematics in Mind,
https://doi.org/10.1007/978-3-319-68246-4_1

also evident whenever we are exposed to beauty in nature and human-made artifacts, and therefore, there is a deep equivalence across human experiences of patterns and order in various and allegedly disconnected domains.

Gaining "general truths" may be conceived as gaining abstract knowledge, which exists in thought but does not have a concrete existence as embodied in matter or actual practice detached from the human mind. How is it possible to abstract "general truths" from the concrete existence in which both human and non-human organisms live their life? How do we learn that two cats, two apples, and two cars all share the mysterious and abstract property of being sets of *two* objects? How do we learn that stretching (without breaking or melding) the dough of a donut changes nothing in the "topology" of the donut, which remains the same? How do we learn the analogy between bee/honey and cow/milk?

Despite the intensive efforts of researchers through the ages, we must modestly admit that the way such knowledge is gained is to a large extent still a mystery. Explanations of the way intelligence emerges and operates are still a way of "substituting one mystery for another" (Mill, 1882, p. 584); they are not yet "explanations" in the most rigorous and scientific sense of the term.

For instance, developmental psychology, such as the one proposed by Jean Piaget, considers the developmental trajectory as moving from concrete to abstract thinking. However, observing young children creating paintings of human beings, one can easily acknowledge the high level of abstraction characterizing the represented figure. It seems, therefore, that there is not a simple linear trajectory leading from the concrete to the abstract, at least not in the minds of human and non-human organisms alike. This argument is strengthened if one examines the linguistic signals of an approaching mental deterioration. Starting to use the phrase "this thing" instead of pointing to a concrete and specific reference is an expression of abstraction, as "thing" doesn't point to any concrete object or concept. However, this expression of abstraction might be a sign of mental decline rather than an indication of healthy abstraction in action. There is no simple trajectory leading from the concrete to the abstract, and, while abstraction has been praised and idealized through the ages, it seems that *learning to attune to the concrete* is no less a challenge.

In sum, the idea of a simple linear path from the concrete to the abstract should be rejected as invalid. Something much more complex and delicate is evident in natural intelligence, something that we cannot easily pinpoint despite our best efforts to provide a simple answer.

The quandary of natural intelligence is also evident when we examine artificial neural networks as a model of human cognition. When these artificial networks "learn," they actually optimize the weights between "neurons" in such a way that the cost or error associated with the cognitive task performed by the network, such as classification (e.g., classifying an object as an apple), is minimized. According to this perspective, natural intelligence could have been shaped through evolution, both biological and cultural, and through personal development in such a way that the (1) *architecture* of the neural network and the (2) *weights* of the neural connections were optimized to meet the cognitive challenges facing the organism. That is, the way in which the neurons are connected (i.e., the architecture) and the weights of activation between the neurons are the two aspects that are mainly responsible for the behavior of the network.

For example, the visual system of many insect pollinators is highly sensitive to ultraviolet light, a fact explained by the need to detect floral signals rich in ultraviolet patterns (Stevens, 2016). According to this explanation, the neural network of such an insect could have evolved in a way that optimizes the identification of sweet and tasty flowers. If you are an insect making your living by consuming pollen, your mission is clear: find flowers, which may provide you with a source of energy. Accordingly, your neural system must be shaped in such a way as to maximize your ability to identify flowers. In other words, and through an evolutionary explanation, the poor insects that were not competent enough at identifying flowers didn't survive, whereas those that somehow possessed better identification devices were lucky enough to survive.

The appealing approach proposed by evolutionary theory faces some serious theoretical and empirical difficulties when it is used to try to explain natural intelligence. For instance, one has to recall that evolution is a "blind watchmaker," to use Richard Dawkins' famous metaphor, a watchmaker that has no interest in teaching or learning. Essentially, in evolution, random variations (e.g., variations in the lengths of birds' wings) subjected to real-world constraints (e.g., advantage to birds with shorter wings in a given ecological niche) result in only some animals moving through the strainer of the environment and propagating (thus creating the next generation), while others are doomed to extinction. However, these mechanisms, whose existence is undeniable, cannot explain all aspects of living systems – not even the most significant of them, such as learning.

For example, *learning by association* is a very general learning principle. Insect pollinators associate the pattern of flowers with rewarding food, a dog associates the whistle of its master with the invitation to jog in the park, and some organisms associate the smell of pheromones with readiness to mate. However, the fact that learning by association is *coded* in our genes, is expressed through intracellular activity, and emerges through interaction with the world cannot be explained by Darwinism or neo-Darwinism. Is there a solid scientific *explanation* for the mechanism through which learning by association has been represented and encoded in our genes, in other words, formed by natural selection?

The idea that "nothing in biology makes sense except in the light of evolution" (Dobzhansky, 1973) represents a general theoretical *stance* that must not be mistakenly confused with the *argument* that (1) everything in biology has an explanation in the light of evolution and (2) points to the exact mechanistic nature of such an explanation. In this context, it is quite difficult to explain how natural intelligence has emerged at the species level of analysis and at the individual developmental level of analysis without adhering to the petitio principii of the Darwinian axioms. Crucially, though, my criticism should not be mistakenly read as a creationist refutation of evolutionary mechanisms but just as a cautious, critical, and contextual understanding of the scope of the neo-Darwinian explanation.

A commonsense point of departure for studying natural intelligence – that is, intelligence as it is observed in natural biological rather than artificial systems – is the idea that natural intelligence is *patterned to the real world*, which means that it emerges, develops, and sustains itself in context. As the world, whether the natural or the social, exhibits some hypothetical order, tangled, noisy, and complex as it may be, natural intelligence should somehow correspond to this order (or orders in the plural). Therefore, a potentially fruitful research plan is to model the *models* of natural intelligence through structures, as structures are simple models that may help us to understand how organisms represent the world.

Identifying structures in human intelligence, social phenomena, and related cultural artifacts (e.g., language) has been the raison d'être of *structuralism*, as illustrated through the work of several eminent scholars from Jean Piaget (1970) to Ferdinand de Saussure (1972, 2006) and Valentine Volosinov (1986). In fact, even the history of the humanities teaches us that since antiquity, and in various cultures, scholars have not been satisfied with descriptive activity only (e.g., describing historical events) but have sought to identify general and abstract structures underlying cultural artifacts from language to literature.

Seeking structures is clearly justified, as the informative value of pure description is doubtful. Think, for example, about the work of a historian who makes the laborious effort to detail the events that led up to World War I. He may weave an enormous web of actors, geographical locations, events, and documents to produce an impressive amount of textual description. However, after being exposed to such a monumental work on which rivers of ink have been spilled, you may end reading this work by asking yourself, "So what? What have I learned about the events that led up to this war, with its painful atrocities? Is there a general lesson that I can learn – one that can be produced through condensing the 'uncompressed' information to which I have been exposed so far? Is there a *general* lesson that I can learn beyond non-informative mountains of information? Is there a general lesson I can learn that is not a cliché?"

These questions cannot be dismissed even if they pull the carpet out from below some of the most established academic disciplines and practices. Holding the belief that the existence of one's discipline and academic practice is justified through its *own existence* is a symptom of moral and scientific decadence. The structuralist movement clearly addressed this danger by trying to move away from mere descriptions and speculations to the identification of general structures that were supposed to be much more informative.

The heroic venture of structuralism, as a general articulated approach to the study of epistemology and human artifacts (e.g., literature), was highly influential for a period of time but for various reasons has since lost its glory. In the humanities and the social sciences, structuralism has been under heavy attack from "post-structuralist" or postmodernist thinkers, who in some cases substituted healthy and necessary philosophical skepticism with malignant cynicism and relativism. As in punk music, the *anti* of postmodernism has often been much more important than its *pro*, which explains why in too many cases this postmodernist agenda has not been constructive beyond the possible importance of its teenagers' rebellion.

On the other hand, the criticism of structuralism was fully justified among those who sought to better clarify the meaning of "structure" and couldn't accept the idea that the complexity of human intelligence, social structures, and various cultural artifacts can be scientifically modeled by reducing them to a set of simple structures. A challenge to the old and simple explanatory schemes proposed by structuralism has recently emerged from the field of *artificial neural networks*. Recent advancements in the engineering of artificial neural networks and in the *deep learning* approach (Goodfellow, Bengio, & Curvill, 2016) have exhibited some remarkable results concerning the performance of intelligent tasks, results that call into question the basic ideas of modeling intelligence proposed (for instance) by Piaget and other eminent psychologists of the structuralist movement.

It must be remembered that structures are theoretical constructs used by researchers to describe the outcomes of highly complex underlying processes in a simple, economical, and communicative way. For example, we observe a child who learns to group objects based on their function rather than visual superficial similarity, and we conceptualize this achievement in terms of the more complex cognitive structure that supports this categorization. However, it is an open question whether formalizing such phenomena in term of structures has any current relevance for understanding natural intelligence in its various expressions. For instance, the way we learn different categories, such as the category of cats or the category of chairs, has been intensively studied by cognitive psychology, which has produced some impressive theorizations. In contrast, researchers have had remarkable success using deep neural networks to categorize visual images with no reference at all to these simple theories and models proposed by cognitive psychology. In many of the success stories of studies in neural networks, the researchers cannot even explain the behavior of the system. The overwhelming success of deep neural networks may raise the question of whether this perspective should lead the modeling of natural intelligence and not the simple cognitive and psychological models of the past.

This is a fundamental question that may gain more and more affirmative answers as time unfolds. One has to be aware, though, that the impressive achievements of these artificial neural network systems are a result of an *engineering* process that has nothing to do *directly* with the modeling and understanding of natural intelligence. An airplane is an incredible engineering achievement that may perform much better in many respects than naturally flying organisms, from the mosquito to the albatross. However, airplanes are tools designed to fulfill a certain function and *not* as models of real flying organisms.

There are, of course, some similarities between artificial and natural flying machines: the engineering of airplanes may be inspired by naturally flying machines, and a biomimetic approach to the engineering of airplanes may even study specific aspects of natural flight in order to build better airplanes. However, an airplane *is not* a model of a bird or any other flying creature. It is basically a flying machine engineered by human beings in order to optimally meet certain aims, from the transportation of people and cargo to airplanes' use as war machines.

In sum, artificial human-made systems, successful as they may be in performing "intelligent" tasks, cannot be trivially used as models of natural phenomena. This clarification doesn't dismiss the possibility that natural intelligence may be studied through the spectacles of artificial machines. After all, the idea of the human brain as a kind of supercomputer able to perform complex computations is the dominant current approach in cognitive sciences. However, considering the brain as a computer, or more accurately as a kind of *abstract* Turing machine, materialized in neurons and synapses, is just an analogy or heuristic, as it is clear that natural computation is quite different from the computation of artificial human-made machines.

As you have probably realized, so far, I have tried to explain the full complexity facing us in trying to understand natural intelligence. This context, in which we are able to critically reflect on the grand theories of the past (e.g., structuralism), doesn't have to discourage us from further studying structures of natural intelligence. There was an appealing sense of security in the grand structuralist theories of the past, which sought simple explanations for what we currently understand as complex processes. In retrospect and from a reflective perspective, we have no choice but to adopt a critical appraisal of these ventures. Impressive as they were, they were naive and too simple.

Given this critical appraisal, the pendulum may easily swing back in the opposite direction, and we may find ourselves dismissing the structuralist agenda while warmly adopting the new and promising path proposed by artificial neural networks. In addition, just as there was an appealing sense of security in the structural theories of the past, there is a contrasting sense of security in dismissing the old theories as totally wrong and adhering to the novel and promising state-of-the-art achievements of present models of neural networks. However, and similarly to the personal growth of a human being, a healthy process of exploration means *distancing from* but not necessarily *rejecting* the secure base from which we have emerged.

Having this metaphor in mind, one possible approach to the study of natural intelligence may seek to describe the deep "mathematical" structures of natural intelligence by rejecting both the naivety of classical structuralism on the one hand and its postmodernist dismissal on the other. In other words, given the assumption that natural intelligence somehow corresponds with real-world patterns, a very minimalist *neo-structuralist* agenda may seek to model the way regularities are represented in the "mind" by introducing structures that model these representations. We may still seek to model by using structures, but structures that pay tribute to the complexity of the mind they seek to model. This neo-structuralist agenda doesn't aim to dismiss or compete with the current approaches to neural networks. In fact, the neo-structuralist approach presented in this book may even explain to us some unsolved issues regarding the architecture of the mammalian brain and therefore may even have relevance for the design of artificial neural networks.

In sum, the aim of this book is to model natural intelligence, specifically that of *human beings*, using abstract structures expressed through the appropriate mathematical formalism. Therefore, the book may be of interest to cognitive scientists, psychologists, philosophers, and researchers in the field of artificial intelligence. Readers must understand, though, that this book is highly theoretical and abstract; I do not aim to develop structures or models of intelligence and to test them empirically but rather to inquire into the very general notion of such structures.

One more qualification should be added for those who still strive to respond to the book through the device of the straw man regardless of the realist and reflective perspective presented thus far. This book has no naive pretentions to propose a new grand and comprehensive theory of intelligence, neither does it offer new recipes for designing better artificial intelligence systems. The book has a much more limited and modest aim, which is to present a *novel, theoretical*, and *challenging* neo-structuralist analysis, or more accurately meta-structuralist analysis, of some aspects of natural intelligence. That is, and following the inspiration of Gregory Bateson, the book assumes that the human mind is patterned to the world in which it is a part and that, therefore, natural intelligence (as the capacity of gaining knowledge of patterns) can be described through abstract structures that may enlighten certain aspects of its activity. Moreover, Bateson (Bateson & Bateson, 1988) has emphasized the relational aspect of this patterning and that in contrast with the reified epistemology that characterized modern Western thinking (Nisbett & Masuda, 2003), our world should not be considered in terms of concrete or abstract objects but in terms of relational patterning. This idea that has fascinated me when I published my first book (Neuman, 2003) has found its expression in the current book where category theory is used as a mathematical language for modeling *relational patterning*. Whether this venture may lead to a better understanding of natural intelligence in its various aspects and whether it may in some way be useful in designing artificial intelligence systems are questions for the future and are beyond the book's main scope. The reader may therefore read this book as an intellectual detective story, where the pleasure accompanying the quest for truth has precedence over all other particularities.

Summary

- Intelligence is about comprehending "general truths" that may be interpreted as patterns.
- Natural intelligence is interwoven with the world and therefore reflects and refracts "orders" existing in the natural, social, and cultural worlds.

- Artificial intelligence is not about modeling natural intelligence. Artificial intelligence is mainly about the development of intelligent tools.
- The structuralist movement seeks to identify "structures" underlying human activity.
- A neo-structuralist agenda should follow this line but also address the failures of past ventures and seek new directions.

Chapter 2
What Is Structure? Piaget's Tour de Force

In the introduction, I argued in favor of a neo-structuralist agenda, but you may have noticed that I have not yet defined or explained what a structure is but have instead used it in the sense of a pattern. We may consider a structure to be an abstract *model* of a relational configuration that optimally supports the ability of the organism to perform various tasks (e.g., recognition). I emphasize the word "model" because a structure is actually a model. We cannot naively assume that structures exist in the organisms' minds and that, for instance, a bacterium moving toward a source of energy (e.g., glucose) holds in its mind a structure that directs its behavior like a small homunculus (or actually bacteriumus) directing its behavior.

A structure is a simple model *we* as researchers use in order to model the modeling process of natural intelligence (this important point should be kept in mind). Therefore, studying structures in general doesn't deal directly with the modeling process as performed by various organisms but with a modeling of modeling (i.e., meta-modeling); this is why in Chap. 1 I described our agenda as "meta-structuralism." For simplicity of discussion, I will sometimes use language that may seem naive in its use of the word "structure" and ignore the nuances presented above, but this allegedly naive use should not be taken at face value.

Let us return to the more basic level of a structure as a model and illustrate it. A human baby, who is born with the ability to identify faces, should be able to recognize the particular faces of his caregivers under various transformations and "distorting" conditions (e.g., lighting).

© Springer International Publishing AG 2017
Y. Neuman, *Mathematical Structures of Natural Intelligence*, Mathematics in Mind,
https://doi.org/10.1007/978-3-319-68246-4_2

A face may be described as a structure because, beyond the particularities of specific faces, it seems that our mind holds a typical relational configuration that may be used as a *guiding template* for the identification of particular and *real* faces; this template may be used to distinguish between human and non-human faces and between faces and other categories (e.g., hands). However, a face is just one instance of a structure. There are mathematical structures (e.g., number), conceptual structures (e.g., the concept of democracy), and musical structures (e.g., those uniquely characterizing Bach). However, it is problematic to assume that the mind holds a library containing numerous structures. It is much more reasonable to assume that the mind is grounded in some basic ability to form structures from which all types of structures grow and adapt in real time. Thus, we should ask whether it is possible to study the concept of structure from a more general perspective than the one evident in its various expressions. The affirmative answer provided by Jean Piaget is discussed in the next section.

Piaget on Structure

In his seminal and insightful book *Structuralism* (1970), Jean Piaget argued that structures in general have certain common and necessary properties. Piaget first considers structure as a system of *transformations* that are responsible for the structure's invariance and hence its identity. Informally, we may regard a transformation as a kind of mapping or translation between two domains, such as in the case of the mathematical concept of permutation, where the elements of a set are rearranged in a certain way. Another example is the translation of a geographical map onto a smaller map by shrinking the distances between the points.

In discussing a structure as a system of transformations, Piaget considers a specific kind of transformation: *structure-preserving transformations*. These transformations are operations we may apply on the structure but operations that preserve its identity. For example, if you have ever baked bread, you know that stretching the dough won't change its identity. Stretching the dough is therefore a structure-preserving transformation. However, and from a mathematical topological perspective, tearing the dough into two separate pieces isn't a structure-preserving transformation. The same idea is evident when you play with graphical software that allows you to process faces. You may stretch a face image horizontally to form a distorted image of the original face. In contrast with the case of the dough, you will notice that at a certain point, stretching the face may change it in such a way that the original image is unrecognizable. We can see that structure-preserving transformations are context dependent: what is true for the dough might not hold for other objects.

The idea of a structure as a system of transformations is very interesting as it suggests that a structure is constituted through some underlying *dynamics* and that it isn't simply a static relational configuration. For example, beyond philosophical complexities, our sense of identity and the experience of a relatively stable and integrated self are considered to be signs of mental health. The "self" may therefore be considered as our way of conceptualizing an organizing psychological structure. We may argue over the question "What is the self?" However, if a person answers the question "Who are you?" by saying "I don't know," we may hypothesize that this person is joking, is engaging in philosophical discourse, or has a deep psychological problem.

The self is primarily a biopsychological concept signifying *boundary-maintaining processes*. The tiger doesn't have to take a course in philosophy in order to intuitively understand that it has a self; otherwise it might prey on itself rather than grass eaters. The same is true for our "immune self" (Cohen, 2000), which is a theoretical structure that aims to explain how the immune system maintains boundaries by tolerating some agents in our body (conceived as belonging to the "self") while attacking others (conceived as "nonself"). The immune "self" isn't a static organization of immune agents but a concept we use to model the complex and heterogeneous dynamic network of immune agents (e.g., T cells) that maintains our integrity by eliminating pathogens, healing damaged tissues, and so on (Cohen, 2000).

Along the same lines, and in the psychological realm, autobiographical memory is a process through which we maintain a sense of psychological integrity – our psychological self – through the reconstruction of life episodes. This is a system of transformations through which past recollections of life episodes are glued together to form an integrated and coherent story of the past as conceived from the first-person perspective: the way "I" see it.

The self as a structure therefore seems to involve a system of structure-preserving transformations that maintain the organism's psychological and biological boundaries and hence its identity. As you may have noticed, this idea seems to entail a bothering circularity as the "structure" of the self involves "structure-preserving transformations." I will try to resolve this point later by adopting a more sophisticated notion of what a structure is.

Piaget's second key idea for characterizing structures in general is that of "wholeness." Wholeness expresses the famous Gestalt slogan that "the whole is different from the sum of its parts" and is not a simple collection or aggregate of elements. A structure is a relational configuration of elements and objects. However, the meaning of the structure cannot be clarified by enumerating its components only. For instance, let's have a look at Smiley's face (Fig. 2.1):

Fig. 2.1 Smiley's face

And then reorganize the two eyes and the mouth as follows (Fig. 2.2):

Fig. 2.2 Smiley's
disorganized face

We have just applied a certain transformation on the components of Smiley's face, but this is *not* a structure-preserving transformation. Therefore, the new organization of the components has turned

Smiley's face into a *non-face*. This is a conclusion we couldn't have reached by enumerating the face elements only. After all, the second representation of the face has the *same* elements albeit organized in a different way. This situation is quite different from the one we observe in simple sets. For example, let's assume that in our world there are only three types of fruits: apple, orange, and cherry. The set of fruits is therefore {apple, orange, cherry}. Now let's change the order of the fruits in the set: {cherry, apple, orange}. This "change" has actually changed nothing in the set as the elements are not connected through some additional relational structure. Therefore, the set of fruits remains identical despite the transformation. However, changing the relational structure of the face elements has changed the *meaning* of the face. We can understand that the unique relational structure of the face elements has formed at the *macro level* of analysis a whole that is different from the simple "sum" or set of its parts. A structure is therefore a set plus "something," and the meaning of this "something" will be clarified as we continue our discussion.

The idea of Smiley's face as eye + eye + mouth + head is not enough to enable us to grasp the unique information conveyed by a face as a whole. Ipso facto, we may understand that we are dealing with a structure if when we strip away an additional layer of organization imposed on a set, we lose the whole. In other words, if we apply a "forgetful" transformation, which drops some or all of the input's structure or properties before mapping it to the output (i.e., the set), then we will lose the unique sense that exists at the macro level. In the case of Smiley's face, applying a forgetful transformation means stripping the face elements of their spatial relational configuration and producing an output that is simply the set comprised of the four elements of the face.

A structure is therefore a whole that is expressed through a unique relational structure based on an underlying system of transformations. Only certain transformations, such as the symmetric reflection of the eyes along the vertical axis, will keep the whole face invariant. This idea of wholeness is crucial as it suggests an epistemological phase transition from the micro-level elements to the macro-level structure.

The idea of wholeness is not unique to the perceptual domain; it is evident also in *semantics* (Neuman, Neuman, & Cohen, 2017), where it may be illustrated through the example of word compounds.

Consider the word compound "hotdog," which is composed of two words: "hot" and "dog." Now, let us assume that you are familiar with the meaning of "hot," as a word signifying a certain high temperature associated with an object, and with the meaning of "dog," as a word denoting a member of the canine family. Understanding the literal and simple meanings of "hot" and "dog" doesn't provide you with the ability to understand the meaning of the *whole* compound word "hotdog" as a kind of sausage that can be eaten.

Let's assume that you have never heard the word "hotdog." You are presented with this word and asked several questions about the object it signifies, such as "Can you eat it?" and "Is it necessarily hot?" Your ability to answer these questions correctly is extremely limited as a result of your lack of understanding of the word compound; your familiarity with the senses of the components does not enable to you give the correct answers. The meaning of the linguistic whole "hotdog" is *different* from the meaning of its components or any simple form of their semantic composition. The example of "hotdog" illustrates a major aspect of wholeness that may be described under the title of *synergy*. Some complex systems involve the formation of macro-level constructions perceived as having features that cannot be reduced to their micro-level constituents. This is the expression of synergy, where the joint action of the constituents produces unique features that are irreducible to the constituents' isolated behaviors or properties or their simple composition. As emerging wholes formed under certain transformations, structures illustrate this synergy, which is a vital aspect for characterizing structures, as will be further discussed.

One should realize that the synergetic aspect of structures may pose a problem for researchers modeling natural intelligence but that it is also an extremely important *solution* for natural intelligence. This is because the unique relational configuration of the structure's components functions as a *constraint* on their potentially enormous combinatorial space. If the whole is different from the sum of its parts, it means that only a limited number of the components' configurations are responsible for the "wholeness" and that one doesn't have to examine the numerous other combinations in order to grasp the whole.

For example, let's assume that our recognition system works along the lines of Bayesian reasoning and that we are supposed to identify x

(e.g., Smiley's face), which is a possible outcome of classes C_x (e.g., face vs. non-face), based on a set of features (y) characterizing the face. The features are the face components. In Bayesian terms, the classification problem is therefore the probability of being a face given a set of features and their organization – that is, $p(C_x|y)$. Again, this formalization signifies the probability of a *face*, in comparison, for instance, with a non-face, given a specific combination of features and their value (i.e., spatial location).

We may also illustrate this Bayesian logic through a natural language processing task. Assume that you are given the task of automatically deciding whether a text expresses depression or not. Your C is therefore comprised of only two classes: depressed and not depressed. In order to automatically analyze each text, you convert it into a vector – an array of words. That is, you convert each text into an ordered list of words and attach to each word some weight indicating its importance in the text. The words in your text are actually the features you use in order to decide whether the text expresses depression or not. To decide whether the text is "depressed," you may use a corpus of texts tagged as depressed and not depressed and calculate the probability of a depressed text given a certain set of features, their value, and their combinations. For example, you may find that given that a text includes words expressing sadness, suicidal ideation, and helplessness, the probability of its being tagged as depressed is ten times more than in the case of non-depressed texts.

Now, let us return to Smiley's example to further elaborate our point and think about the potential combinations of ways in which we may organize the face components: the two eyes may be located beneath the mouth, the nose may be located to the left of the mouth, and so on. This *combinatorial space* of features (i.e., the face components) might be quite big, and so learning to classify C_x as a face given all possible configurations of the components might overload the system and impede its learning and reasoning processes. The fact that the face exhibits only a *limited* subset of this combinatorial complexity reduces the cognitive load associated with learning and recognition/classification.

Moreover, the "symmetry" of the face components, which is an expression of the constraints imposed on the components' organization, makes it easier to evaluate $p(C_x|y)$ using $p(y|C_x)$ (i.e., the prediction problem), as follows:

$$p(C_x|y) = \frac{p(C_x)p(y|C_x)}{p(y)}$$

In sum, while the synergetic wholeness of structures is quite a challenge to model, it is actually a central and cognitively economical aspect of natural intelligence. The fact that the whole is different from the sum of its parts reflects both the organized structure of the world and the way in which natural intelligence represents, constructs, and reconstructs this order. We can therefore see that great importance is given to symmetry, which is a significant factor in the economical computation of categories (Lin & Tegmark, 2016). Wholeness can be decomposed into symmetric parts (i.e., local symmetries), and the meaning of such a local symmetry, which may be quite different from the simple symmetry we observe in nature, will be a cornerstone of the thesis that I develop.

The third basic property of a structure proposed by Piaget is that of *self-regulation*. A structure is a closed system in the sense that its boundaries are preserved under its constituting structure-preserving transformations, from *within*. In other words, a structure can be subjected to various transformations, such as when we slice a piece of dough into two parts. Not all of these transformations are structure preserving, including ones that change the topological structure of the dough. However, to maintain its "closure" – that is, to maintain a *boundary*, which is a characteristic of wholeness – a structure should include some kind of built-in self-regulating mechanism.

In biological systems, self-regulation is achieved through feedback loops that keep the system from losing its boundaries. For example, the regulation of our body's temperature is a task carried out *within* our body and as an in-built maintenance function of the whole body. In contrast with cars, which have an artificial thermostat to regulate the engine's temperature, the human body has no thermostat regulating its temperature from the outside. The body *self*-regulates its own temperature through a neural feedback system mainly associated with the hypothalamus in the brain. Thermoregulation in the human body is therefore self-regulated.

Integrating Piaget's three main tenets of structure, we may consider structure as a relational configuration of elements generated and maintained through a system of transformations and their self-regulating function, where on the macro level, there are novel properties and behavior that cannot be reduced to the micro-level components.

This is a tentative consideration of structure, and the picture will become much more complex and hopefully more interesting as the text unfolds. However, after proffering this promise of further complexity, we may ask whether it is possible to express these properties of structure in more basic and abstract language. This is a critical question for those seeking to model structure across domains. For example, a physicist studying the escape behavior of a crowd during a fire in a mall may use precisely the same model as a physicist studying the movement of gas particles. The same may be true for the structuralist, who seeks to use the same general idea of structure whether he is studying the structures underlying child intelligence, the structures underlying different languages, or the beauty expressed in architecture. It is an open question, of course, whether seeking a general structure is a constructive strategy, but for Piaget the answer was positive.

In trying to formulate the notion of structure, Piaget (1970) used *group theory*. Group theory is a field that studies algebraic structures, which are *sets* and some *operations* defined over these sets. This idea will be made clear subsequently, but, if you recall the example of Smiley's face and the idea of a forgetful transformation, then you may start making the link in your mind by thinking about the possible operations we may apply to Smiley's components.

There are various algebras and various algebraic structures. Group theory studies a specific algebraic structure known as a *group*. A mathematical group (G) is an algebraic structure consisting of a set of elements and a binary operation (*) on G – that is, an operation on an ordered pair of elements, such as in the case of addition operating on a pair of natural numbers, as in the example $2 + 3 = 5$. In this case, the binary operation * takes the form of the addition operation +.

The group structure has several properties, and here we discuss only three of them. First is *closure*. The idea is that using the binary operation keeps us within the boundary of the system. For instance, *integers* form a closed system with regard to the operation of addition.

Adding two integers will result in an integer. Similarly, reflection along the vertical axis of the nose will leave the perception of a face relatively the same. However, this is not true for inversion, indicating that not all transformations result in closure.

The second property of a group is the existence of an *identity* element, which means that combining this element with each element of the set will not change its value. For example, the identity element of integers with respect to addition is 0. Adding 0 to an integer will result in the integer (e.g., $1 + 0 = 1$). It won't change the integer's value. We can see that the existence of an identity element is deeply associated with the property of self-regulation and closure, as presented before. The identity element keeps us within the boundaries of the system and maintains the structure's closure.

A third, and a highly important, property for the thesis developed by Piaget is the existence of an *inverse* element (a^{-1}) that in combination with another element yields the identity, or neutral, element. For example, the inverse of the integers with respect to addition is $-a$ such that $a + (-a) = 0$ (e.g. $5 + (-5) = 0$). Again, the inverse in combination with the identity guarantees the system's closure and hence the boundary of wholeness.

When he describes the algebraic structure of a group as *a prototype of a structure*, Piaget is making an interesting comment, saying that the group represents a unique form of abstraction – *reflective abstraction* – "which does not derive properties from *things* but from our ways of *acting on things*, the *operations* we performed on them" (Piaget, 1970, p. 19). Reflective abstraction is one of the most complex and interesting concepts in Piaget's theory, so let us try to better understand what he has to say about it.

A simple form of abstraction, argues Piaget, involves the extraction of a general property from a thing or a set of things. Let's take the set of fruits as our example. Given the set we have used before – {apple, orange, cherry} – we may extract from this set a general property that may be used to define the set not by enumerating its elements (an *extensional definition*) but by describing the general property shared by all elements of the set (an *intensional definition*). For example, we might say that a fruit is a seed-bearing plant. The intensional definition is a clear form of abstraction as it describes the general characteristic of the set's elements.

Piaget is saying that this form of abstraction definitely tells us something about the thing that is being defined. However, the more general this abstract property, the less useful it usually is. We may understand this point if we describe this form of abstraction as squeezing some general properties from members of a set – for example, cherries are described as sweet, and crayfish and elephants may both be clustered as animals in hierarchical semantic taxonomy (e.g., WordNet). Describing cherries as "sweet" is a kind of abstraction, but many other fruits may be "sweet" and so are other types of food, from cakes to liquors. Saying that cherries are sweet is a powerful form of abstraction, Piaget would have argued, but at a certain point, it might lose its informative power as it may cover too many things. *Overgeneralization* is just one possible consequence of such abstraction.

In contrast with the intensional form of abstraction, reflective abstraction involves the operations we may apply to an object – for instance, the structure-preserving transformations of the dough. However, my interpretation is that we may extend this notion of abstraction to include the *abstraction of transformations*, which results in a higher-level identification of generalities. For example, a child may jealously notice that the piece of cake she has been given is *smaller* than the piece of cake held by her sister. She may also realize that her sister is *taller* than her. In both cases, the child is aware of the existence of two single dimensions: height and size. When projected into a higher level of analysis and reflected upon, the one-dimensional relations of "taller" and "bigger" may both be abstracted into the new concept of *order*. Regardless of the difference between the conceived objects (i.e., cakes and people), the *similarity of the differences* between the bigger and smaller cake, and between the taller and the shorter sister, is a similarity of relations that once realized gives birth to the more abstract concept of *order*. I believe that Piaget's idea of reflective abstraction, as interpreted and extended above, cannot be separated from the understanding of what a structure is. The idea of structure is formed through reflective abstraction and the abstraction of transformations. Let's keep this key idea in mind while moving forward. We will return to it later.

I have previously explained that, for Piaget, the three basic properties of structures are transformations, wholeness, and self-regulation.

Moreover, these basic properties may potentially be modeled through the mathematical concept of the group. Therefore, it seems at first that the idea of structure may be perfectly modeled through the mathematical concept of a group. However, here comes the problem. For Piaget, *a crucial property for a structure is reversibility*. Think, for instance, about an object's invariance under spatial transformations. For Piaget, our ability to rotate an object (in mental imagery) and return it back to its original position is crucial for grasping its invariance and hence its perceived structure. A child who rotates an object may learn that the object remains the same despite these rotations. When grown, the child may use the same abstraction in order to understand the idea of energy preservation. For example, she may learn in high school that the energy carried by food is transformed into the energy that moves her body during physical activity. Energy is preserved and never lost.

Piaget further argues, and from a more general and abstract perspective, that *reversibility results in self-regulation of the system* and in the *constitution of the system's boundaries*. To explain these assertions, Piaget insightfully points out that a binary operation is reversible in the sense that it has an *inverse* (1970, p. 15) because an erroneous result is "simply not an element of the system (if $+ n - n \neq$ 0 then $n \neq n$)" (1970, p. 15) (which is a contradiction). This is an insightful observation that deserves further elaboration.

To recall, in group theory, the principle of the inverse means the existence of an element that combined with any other element results in the identity element. For instance, given the operation of addition, the inverse element of each integer a is $-a$. It means that, when adding -3 to 3, we will get 0, which is the identity element, as $3 + 0 = 3$. What happens if we violate this "logic of the inverse"? What would happen if $-3 + 3$ was *different* from 0? The result would be the violation of the *law of identity*, which states that a thing is identical with itself. Violating the law of identity might have catastrophic implications, as identity, which is the most basic aspect of certainty, would have vanished. In a world where the most basic law of identity is violated, nothing can be thought, said, or acted upon. Therefore and according to Piaget, *the existence of the inverse is what allows the formation of the structure's boundaries* as an epistemological and a cognitive must.

In sum, and at least according to Piaget's abstract formulation of structure, reversibility, which is a property emerging from the existence of the inverse function, is a necessary property of structure as it guarantees the law of identity. However, Piaget painfully realized that biological and cognitive systems alike are mostly irreversible! Let me explain this idea and how it challenges the idea of identity and structure.

Reversibility and Irreversibility

The idea that biological and cognitive systems are mostly irreversible, as acknowledged by Piaget, is deeply grounded in the insights gained from the physics of computation (Bennett & Landauer, 1985). A process of computation, in the most general sense of the term, is a process in which some output is produced from some input through certain operations. The addition of two integers, for instance, is a process of computation. The permutation of Smiley's face is a process of computation. Categorizing an object as an "apple" in our mind is a process of computation.

A process of computation is defined as reversible if the input can be restored from the output. For example, the NOT operation in logic can be applied either to the value of TRUE or to the value of FALSE. When we apply the NOT operation to the input value TRUE, we get the output value FALSE, and when we apply the NOT operation to the input value FALSE, we get the output value TRUE. When we get the output FALSE knowing that the operation NOT has been applied, we can be *certain* that the input was TRUE and vice versa. NOT is therefore a reversible operation.

If you think about it from a philosophical perspective, you may realize that NOT is a time-invariant operation and may wonder whether you may imagine a single case where it exists in nature. Later (and, surprisingly, through a short paper written by Freud), we will see how the logical operation NOT is grounded in the mental act of *negation*, which has some interesting aspects.

The NOT is a reversible logical operation that exists in the artificial realm of logic. How about the operation of addition and the output 10? Is there a way of reproducing the input that produced this output? 10 can be the output of 5 + 5. However, it can also be the output of 9 + 1, 8 + 2, and so on. Therefore, the process is *irreversible*. While given the output FALSE and the operation NOT, we are certain to conclude that the input was TRUE; this is not the case in the addition example and in many other cases of natural computation.

The physics of computation explains that during a process of irreversible computation, some information, which is described in terms of *differences*, is *necessarily* lost unless we keep track of all recorded inputs, which implies an almost impossible burden on memory (Bennett & Landauer, 1985).

Imagine, for example, that you hold in your hands two identical elastic rubber balls. In your right hand, you hold the ball 1.5 m above the ground. In your left hand, you hold the ball 0.5 m above the ground. Now you drop the two balls from two different heights at the same time, and the balls fall to the ground. When they hit the ground, they bounce, and the height of the bounce is indicative of the height from which they were dropped. As time unfolds, and as a result of friction with the ground, the balls exhibit less and less information about the height from which they were dropped. The *differences* in height from which the balls were dropped vanish when the balls finally rest on the ground after exhausting all of their energy. This is an example of computation in the physical realm, but you may also think about irreversible processes in the context of natural intelligence.

Imagine a case where you "compute" the intensional meaning of fruits as seed-bearing structures. If you have naturally learned this sense, you have done it by experiencing many instances of fruits. Each apple you have encountered is unique, exhibiting a difference from apples you have encountered before, but you haven't kept track of each and every fruit you have come across in your life. The differences have been totally lost in favor of a more general, abstract, and economical abstraction that guides your behavior.

The physics of computation also explains that there is a minimal energetic price we must pay to apply an irreversible operation, a price evident in the release of entropy into the environment. This idea may be explained if we understand irreversibility in the context of the second law of thermodynamics. The law suggests that the total entropy of an isolated system increases over time. Cognitive and biological systems are clearly *not* isolated systems as they exchange energy and matter with their environment. However, from a probabilistic perspective per se, one can easily understand the law.

As time unfolds, any given system that is left on its own will exhibit an increasingly disorganized configuration of its elements, as this form of organization, actually of disorganization, is the most probable. For example, the mysterious property of being alive involves the active orchestration of biological agents (e.g., cells), organs, and processes in a way that maintains organization and function (i.e., the organism), far from equilibrium. However, when an organism stops living, it is *dis*organized. Hence, death involves a process of disorganization or decay, which is expressed in the defragmentation of the living organism and the return to the more probable state: ashes to ashes, dust to dust.

Ipso facto, one may infer that life somehow overcomes the second law. However, the demon of the second law cannot be pleased but only tricked (Cohen, 2017) in the sense that a local and temporal increase in order, such as the one evident when computing, is somehow compensated by a loss of information and an increase in entropy in the environment. This logic proposed by the physics of computation is applicable to natural intelligence too, as when organisms reach "general truths," as a result of computation; this process is *necessarily* accompanied through the loss of information that exists at a lower level of organization. Think, for example, about the formation of *semantic memory*, which describes the general knowledge that we have acquired through life. A child who grew up in the United States during the 1950s watching the short-film animations of Tom the cat and Jerry the mouse will have acquired the knowledge that *Cat chases after Mice* regardless of his real experience with cats and mice. This factual "declarative" piece of knowledge that Cat chases after Mice has therefore become a component of the child's semantic memory and his represented meaning of Cat.

Cat could also have been understood as that which chases after Mice from a relational perspective. This knowledge, though, is an abstraction that has been formed by *forgetting* a huge amount of information dealing with the particularities of cats, mice, and their interactions. Whether Tom is chasing Jerry in the house or in the yard is of a minor importance; whether Tom is chasing Jerry or another mouse, or whether another cat is chasing another mouse, is of minor importance. The semantic memory is formed by throwing away many particularities in favor of new and sometimes more important generalities. If our child grew up to be a miller owning a gristmill, he probably encountered the problem of mice eating the grains. At this point, he may have considered bringing a cat to the gristmill, recalling that "cat chases after mice." The particular aspects of cats are of less importance. In sum, it seems that natural cognitive systems are to a large extent *irreversible*, a fact deeply grounded in the physics of computation and a fact with detrimental consequences for modeling natural intelligence and structure through mathematical models, which are reversible by definition.

As human beings, we think through brains, which are biological organs grounded and constrained by physical processes obeying certain laws. Piaget chose the mathematical structure of the group for modeling structures in general as he conceived the group "as a *kind of prototype of structures in general*" (Piaget, 1970, p. 19, my emphasis). However, realizing that irreversibility pervades natural intelligence might pull the carpet from underlying Piaget's appealing thesis, as Piaget himself admitted. After all, the group as an abstract mathematical structure necessarily includes reversibility, which seems to be in contrast with the basic logic of life.

Piaget may have failed in modeling structure by adopting a very restrictive mathematical notion of structure that includes limiting notions of identity and reversibility. However, he seemed to point in an interesting and fruitful direction. Is there a way of saving Piaget's thesis? In a paper that I published several years ago (Neuman, 2013), I positively answered the above question. However, as we have learned before, there is no gain without some loss, and in trying to save Piaget's structuralist agenda, we have to sacrifice some of his naive mathematical pretensions. According to my proposal, Piaget was right in pointing to the role of reversibility and the inverse function in the formation of structures, but he failed in proposing the relevant mathematical formalization that may be used to model these processes in the context of natural intelligence.

As a first step in saving Piaget's structuralist agenda, we may therefore want to adopt the idea of natural intelligence as materialized in the general architecture of objects and relations, such as that expressed in the network of neurons or in the conceptual network modeling semantic memory. Therefore, we may ask whether one may formally identify the meaning of reversibility and the inverse function in basic relational structures. Second, we may seek a mathematical formalism that is less restrictive in its demands – one that allows us to substitute, for example, "identity" with "similarity" or "structural equivalence." In using such a formalism, we may be able to locate these "softer" versions of structure in a more general, noisy, and messy context of natural intelligence, in which a delicate balance exists between the loss of information on one level of analysis and the gain of information at another level of analysis. To address this challenge, I have used category theory, which is introduced and explained in the next chapter. My use of category theory is as a modeling language only and implies no pretensions to expertise in category theory as a mathematical field.

Summary

- A structure is an abstract model of a relational dynamic configuration.
- For Piaget, a structure involves a system of structure-preserving transformations, which maintain their "wholeness" through self-regulation activity.
- Piaget used group theory to model structure, as he conceived the group to be a "prototype" of a structure.
- However, the mathematical concept of group involves reversibility, whereas natural computational processes are mainly irreversible.
- Addressing the irreversibility issue while staying within the scope of Piaget's main ideas is a challenge to be addressed.

Chapter 3
Category Theory: Toward a Relational Epistemology

Category theory is a mathematical formalism that describes many similar phenomena across various mathematical fields (e.g., Adámek, Herrlich, & Strecker 1990; Goldblatt, 1979; Lawvere & Schanuel, 2000). It is allegedly very simple since it deals with objects and maps between those objects (denoted by arrows). Its ability to do this, however, makes it a general and powerful language for modeling beyond mathematics, specifically in fields in which mappings and transformations take place at various levels of abstraction. Hopefully you will now easily understand why I've chosen to use category theory as a modeling language, despite the fact that, excluding a few cases (e.g., Ehresmann & Vanbremeersch, 2007; Ellerman, 1988; Phillips & Wilson, 2010; Rosen 2005), it has rarely been used in the social or cognitive sciences or in the humanities. My main inspiration (and major reference) for studying category theory is *Conceptual Mathematics* (Lawvere & Schanuel, 2000), although additional references will be used and cited.

The first type of component in a category is objects (A, B, C, etc.). A category always includes some objects/components, such as those included in a set. However, we will shortly realize that a category may also be defined (perhaps less intuitively) in terms of its more dynamic aspect of transformation, expressed as *maps* between objects. It is important to keep this idea in mind as it contradicts some of our epistemological prejudices. The second type of component in a category is therefore *maps* between objects (denoted as *f*, *g*, *h*, etc.). These maps, or *morphisms*, graphically represented as arrows, are merely a

© Springer International Publishing AG 2017
Y. Neuman, *Mathematical Structures of Natural Intelligence*, Mathematics in Mind,
https://doi.org/10.1007/978-3-319-68246-4_3

Fig. 3.1 A map illustrating
the permutation of three
boxes

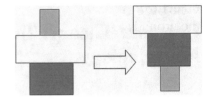

way of relating the objects to themselves and to each other. These
relating maps may be considered as the transformations discussed by
Piaget in the context of structure. Morphisms/maps can be used to
model natural processes. If we model the category of cortical networks, for example, the objects may correspond to neurons and the
morphisms to relations of activation between these neurons. If we are
modeling natural language, objects can be used to represent nouns
and morphisms to represent verbs.

For each map (symbolized for illustration by f), there is one object
that functions as the *domain* of f, the source from which the arrow
originates, and another object that functions as the *co-domain* of f,
the target object where it ends. For example, map f from domain A to
co-domain B is represented as an arrow from A to B [$f: A \rightarrow B$]. In
Fig. 3.1, you can see on the left a set of three boxes and on the right
the same set of boxes differently organized. The arrow from the first
organization (i.e., the domain) to the second organization (i.e.,
the co-domain) symbolizes the morphism that permutes the boxes.
The set of boxes plus the permuting morphism define a category.

At this point, it is critical to explain that mapping can take place at
a higher level of abstraction *between categories* and not only between
objects. *Functor* is a term denoting the map from one category to
another, and we will delve deeper into this important idea as we
proceed.

There are two additional properties of a category that I would like
to introduce. The first one is an *identity map*. For each object A in a
category, there is an identity map – 1_A – that has a domain A and a
co-domain A: $1_A: A \rightarrow A$.

If you think about the existence of an identity map from a broad
philosophical perspective, you may be surprised to realize that the
identity of an object may not necessarily be defined in *essentialist*
terms, through some kind of immanent "essence" or property characterizing the object, but through a transformation leading from the

object to *itself*. This idea may be dismissed as logical circularity on the grounds that the object is defined through a map to itself, the same "itself" that is defined through a map to itself, and so on ad infinitum.

However, the nontrivial meaning of this alleged circularity may be illustrated through a scene from *Alice's Adventures in Wonderland* (Carroll, 1865/2009), where Alice encounters the Caterpillar sitting on a mushroom and smoking a hookah. After their short conversation, the Caterpillar is transformed into a butterfly. Is this transformation an identity map from the object Caterpillar to itself? On the one hand, it is clear that the Caterpillar isn't a Caterpillar anymore but a new thing we call Butterfly. On the other hand, assuming that the newborn Butterfly is a transformation of the same cogent creature that appeared to us as the Caterpillar, a cogent creature with consciousness and autobiographical memory, then we cannot object to the idea that the transformation is an identity map. The mysterious object Alice has witnessed in transformation is an object defined by the identity map regardless of its different appearances at successive time points. However, we may question whether this transformation is an identity map or one of the most famous types of mapping, known as *isomorphism*.

The map f: A → B, is an isomorphism if there is a map g: B → A, for which the symbol "∘" stands for "following" and $g \circ f = 1_A$ and $f \circ g = 1_B$. In this context, A may stand for the Caterpillar and B for the Butterfly. From an irreversibility point of view, the transformation from the Caterpillar to the Butterfly is an isomorphism if it can be reversed, and the Butterfly can turn back into a Caterpillar. This doesn't seem to be the case; although in *Alice's Adventures in Wonderland*, there are some transformations that clearly express isomorphism. Turning into a Butterfly, though, seems to be an irreversible process, as in the real world (as opposed to in its imaginary counterpart) we can't reverse the flow of time.

The amusing and quite bizarre scene taken from *Alice's Adventures* may help us to better understand the nontrivial aspect of the identity map and even the nontrivial sense of isomorphism expressing the highly important property of *reversibility*. Isomorphism is another central notion that must be kept in mind as it is crucial for the thesis developed in this book. At this point we may equate reversibility with isomorphism; however, as we proceed through this book and lose the notion of isomorphism in favor of structural equivalence, we may start thinking about reversibility in more complex terms.

The next property of category that I would like to discuss is that for each pair of maps h: A → B and g: B → C, there is a *composite map* e: A → C (in other words, map $e = g \circ h$, where ○ means "following"). What does it mean that we have a composite map? This idea may seem to be strange from a psychological perspective, as illustrated through a simple example. Let's assume that A stands for Abigail, B for Bernard, and C for Charles and that each map stands for "love." Therefore, Abigail loves Bernard and Bernard loves Charles. Can we form a meaningful composite map implying that Abigail loves Charles? There is no logical necessity or even empirical probability that Abigail loves Charles. She may love Charles, hate him, or be indifferent to him. However, the idea of a composite map just says that if two paths originate from the *same* source and end at the *same* target, then they are the same, as what is important is the output we have produced from some input. In other words, the idea of composition is used in a very simple sense and doesn't imply transitivity.

Given the characteristics of a category, we may now understand that mapping can take place between categories. To recall, the term functor is used to denote such a map. If we have two categories C and D, then a map from C to D (i.e., a functor) associates with each object in C an object in D and associates with each morphism in C a morphism in D in a way that preserves identity and composition. This is a key idea as mapping between categories is not a simple mapping of objects from one category to the other. A functor is a kind of map that respects the *internal structure* of each category as it is restricted by the mapping of morphisms preserving both identity and composition.

Up to now I have shown that mapping is formed between objects, which may lead to the erroneous impression that in category theory objects have an ontological and epistemological precedence over morphisms. From a simple perspective, categories are collections of objects (i.e., sets) plus morphisms that impose on these objects some structure. However, in category theory, morphisms have *precedence* over objects, and (surprisingly as it may seem) a category may even be defined without using the notion of objects. To explain this counterintuitive idea, let me introduce several concepts. A *partial function* from X to Y is a function X′ → Y for some subset of X. It is not obligatory to map every element of X to an element in Y. If we are discussing a binary function that combines two elements of a set to produce another element of a set and if this binary function or operation is partial, then we call it a *partial binary function*. A partial binary algebra is $(X, *)$, where X is a class (i.e., a large collection of sets), $*$ is a partial binary operation, and an element of X is called a unit of $(X, *)$.

Now to the object-free definition of a category (Adámek, Herrlich & Strecker, 1990). An object-free category is a partial binary algebra $C = (M, \circ)$, where the members of M are called morphisms and the sign $\circ$ stands for composition. You see, what we have is just morphisms and their composition. This algebra satisfies the following conditions:

1. Matching condition: for morphisms f, g, and h, the following conditions are equivalent:

    (a) $g \circ f$ and $h \circ g$ are defined.
    (b) $h \circ (g \circ f)$ is defined,
    (c) $(h \circ g) \circ f$ is defined.

2. Associativity condition: if morphisms f, g, and h satisfy the matching conditions, then $h \circ (g \circ f) = (h \circ g) \circ f$
3. Unit existence condition: for every morphism f, there exist units u_C and u_D of $(M, \circ)$ such that $u_C \circ f$ and $f \circ u_D$ are defined.
4. Smallness condition: for any pair of units (u_1, u_2) of $(M, \circ)$, the class $\mathrm{hom}(u_1, u_2) = \{f \in M | f \circ u_1 \text{ and } u_2 \circ f \text{ are defined}\}$ is a set.

In sum, this definition proposes that a category is basically a class of morphisms. This object-free definition of a category is much more complex and abstract than the object-based definition. However, tricky as it may seem, it gives us the possibility of formalizing a category in terms where morphisms have precedence over objects. I believe that the importance of this object-free definition for modeling natural intelligence is in directing us to adopt a *dynamic perspective* on the realm we are striving to model. Instead of populating our reality with ready-made objects, an approach that has some problematic consequences for understanding in epistemology and in the cognitive and social sciences (e.g., Neuman, 2003), we should focus our efforts on modeling the transformations that give rise to structures of interest. This is highly important if we are moving along Piagetian lines, as reflective abstraction is the abstraction of transformations and not simply the abstraction of objects. Given these basics of category theory and their minimally elaborated sense, we may turn to "universal" structures identified through category theory.

Universal Constructions

Despite its apparent simplicity, a category is a starting point that gives rise to incredibly abstract and complex structures. Following Piaget, the main objective of the following paragraphs is first to find *an analogy for the inverse element in category theory* and then, by using this analogy, to (1) investigate structure in categorical terms and (2) use this formalization of structure to model natural intelligence that is basically irreversible. We will follow this plan step by step, as identifying the inverse is crucial for establishing reversibility, which is necessary for the formation of structure. To achieve this aim, several basic definitions first need to be introduced.

We start with the definition of an *initial object*. An object 0 is initial in category C if for every C-object *a* there is one and only one arrow from 0 to *a* in C (Goldblatt, 1979, p. 43). If we think of a category as a directed graph composed of nodes (or vertices) and arrows between the nodes, the initial object is the node that sends one and only one arrow to *each* of the other nodes. It is important to realize that arrows may point to the initial object. That is, by definition, arrows go from the initial object to all other objects in the category, but arrows may also reach the initial object from other objects in the category.

An illustrative example from natural language may clarify the meaning of the initial object. Think about the first-person pronoun (singular) "I." "I" functions as a kind of initial object as it is always the source of "arrows" to other objects in its category: "I like cats," "I enjoy swimming," etc. The "I" may operate on various objects, but "I" is *never* the co-domain or target of another object. When the "I" turns into a target or co-domain (i.e., when it turns into an object of a transitive verb), it is given another linguistic sign, which is "me" or "myself" (e.g., "He told me …"). The characterization of "I" as an initial object may explain its unique status as a linguistic sign. Mikhail Bakhtin, who was an insightful epistemologist, has reflected on the unique status of the "I" (Bakhtin, 1990), saying that the "I" is the only sign in language that has no clear designatum. When I use the word "cow" in its literal sense, I signify a certain set of animals. When I say "what a beautiful bird," I use the word "bird" to designate a specific bird flying in my garden. However, when I use the sign "I," it doesn't point to any specific object or concept. Bakhtin's wonder about the "I's" unique status may be explained through the status of the "I" as an initial point and as the semiotic node of our first-person perspective.

A complementary notion to the initial object is that of the *terminal* object. An object 1 is terminal in category C if for every C-object a there is one and only one arrow from a to 1 in C (Goldblatt, 1979, p. 44). Along the lines of the previous example, the pronoun "her" is a terminal object as, in the category of signs in which "her" is weaved, the "her" will always be the object toward which arrows are directed. Note that any two initial or terminal C-objects *must be isomorphic* in C. The proof goes like this. Assume that T_1 and T_2 are two terminal objects in a category C. By definition, there is exactly one map from each object in C to the terminal object. Therefore:

$$T_1 \rightarrow T_2 \rightarrow T_1$$
$$T_2 \rightarrow T_1 \rightarrow T_2$$

and T_1 and T_2 are therefore isomorphic.

At this point, it is important to know that isomorphism means that the objects are *indistinguishable* (Goldblatt, 1979) and hence *exchangeable*. If you cannot distinguish between two objects then for all practical intents and purposes, they are exchangeable. This is an interesting point, as, following one of the greatest structuralist thinkers, Ferdinand de Saussure, the concept of *value*, which Saussure (2006) equated with *meaning*, is grounded in exchange, which may be formalized up to isomorphism.

For example, let's take the field of economics. One US dollar has value as long as it can be exchanged for material objects (e.g., a beer opener) or other coins such as the Swiss franc. Money is valuable as long as it can be exchanged, and it can be exchanged as it has a value. According to this semiotic perspective, a sign has meaning as long as it may be exchanged for a concept or action. The sign "sit!" has meaning as long as Snoopy the dog may interpret it as an order to sit. For a cat, and regardless of any efforts, this command is meaningless. And now let's return to initial and terminal objects.

Saying that all initial and terminal objects in a category C are isomorphic means that they are in a sense the same and exchangeable, and therefore have the same meaning or value. Now, Fig. 3.2 describes a structure in which all three objects are isomorphic:

Fig. 3.2 A structure with three isomorphic objects

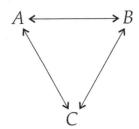

In the structure shown in Fig. 3.2, A and B are isomorphic, B and C are isomorphic, and A and C are isomorphic. The objects A, B, and C are isomorphic and therefore indistinguishable. This means that none of the objects has a distinguished identity and, furthermore, that we have no way of identifying which is A, which is B, and which is C except through their arbitrary tagging as such. In other words, as each of the objects has exactly two bi-directional arrows pointing to and from it, knowing which is which by examining the arrows or connections would lead us nowhere and for all practical purposes we may regard the three objects as undistinguishable – they have no essential "internal" identity. At this point you may understand why tagging or "naming" is a key cognitive activity when trying to elucidate the meaning of a structure. The above tags – A, B, and C – are arbitrary precisely because we could have replaced them with other tags with no implication whatsoever for differentiating the triangle's nodes. Mapping the triangle's nodes to names is therefore a powerful conceptual tool.

A similar idea can be found in semiotics. As proposed by Saussure, in any sign system the signs are differences only and have no essential identity. The sign "dog" has no essential meaning. Starting from now, we may replace the word "dog" in English with the string of symbols "#C3Et." Nothing would have changed if this new norm was to be accepted. Signs, according to Saussure, have meaning only by being differentiated from each other and related to each other and by corresponding with the conceptual and material realm they signify. The lack of distinction between the nodes in Fig. 3.2 means that the graph is perfectly symmetric and thus that exchanging one node for another would change nothing in the graph as a whole.

Isomorphism is an *equivalence relation* that may be used to divide a category into sub-classes of isomorphic objects. The structure in Fig. 3.2 is indivisible as all of its objects are isomorphic. In what sense is this "whole" different from the sum of its parts? You have noticed that the above structure is actually a triangle. A triangle is defined as a shape with three edges and three vertices. However, knowing about the components of a triangle probably will not help us to conclude that the sum of a triangle's angles is equal to 180°. The whole presents an emergent property that cannot be deduced from the sum of its parts. We will get to this point later when discussing wholeness and Gestalt structures.

The next concept that I would like to introduce is the *product* (Goldblatt, 1979, p. 47). This is an important concept as it is a building block of a hierarchical system. Given two objects A and B, the product of A and B is (1) an object denoted A × B and (2) two maps called *projections* $(pr_A : A \times B \to A, pr_B : A \times B \to B)$ (see Fig. 3.3):

Fig. 3.3 Projections of A × B
$$A \xleftarrow[pr_A]{} A \times B \xrightarrow[pr_B]{} B$$

For any other object C and any maps $f : C \to A$, $g : C \to B$. There is exactly one arrow $\langle f, g \rangle : C \to A \times B$ such that following diagram in Fig. 3.4 *commutes*, meaning that $f = pr_A \circ \langle f, g \rangle$ and $g = pr_B \circ \langle f, g \rangle$:

Fig. 3.4 The commuting diagram

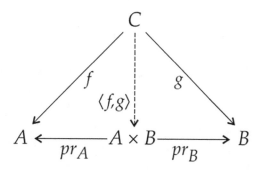

The "universal" property of the product is such that, if there is an object C that can be mapped to A and B, then it *must* factor through A × B. In this light, it is clear why the product is described by Lawvere

Fig. 3.5 The taxonomy of Chair

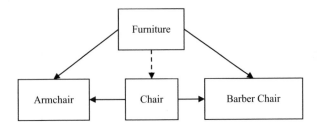

and Schanuel (2000, p. 269) as the "best thing of its type." This means that every other object equipped with maps to A and B must have *exactly one* map that makes the diagram commute; that is, we must have a *unique* path through which to pass in order that $f = pr_A{}^\circ \langle f, g \rangle$ and $g = pr_B{}^\circ \langle f, g \rangle$.

The idea of the product as the "best thing of its type" is very interesting and deserves further elaboration. For instance, let's assume the existence of a taxonomy in which we have Chair, its *hypernym* Furniture, and its *hyponyms* Armchair and Barber Chair. A Chair is an item of Furniture and Armchair and Barber Chair are Chairs. Therefore an Armchair and a Barber Chair are also Furniture. These taxonomic relations are represented in Fig. 3.5.

As you can see, the path from Furniture to Chair is unique (i.e., a chair is a Furniture), and the diagram commutes in the sense that it represents the hierarchical structure of this semantic taxonomy and expresses the *universal construct* of the product. The Chair may be considered as the "best" thing of its type in the sense that it provides us with the *optimal* conceptual level for describing a thing. Indeed, when shown a picture of a Poodle, one would seldom describe it as an Animal, as this concept is too general to be informative. The informativeness of Dog in contrast with the too general concept of Animal is explainable through the universality of the product. On the other hand, giving a more detailed specification might be too informative and violate the idea of information relevance. A Poodle will usually be described through the concept of Dog, which is the best thing of its type.

As Lawvere and Schanuel (2000) explain, refuting the claim that our product is the "best thing of its type" would simply require that we show that there is another product object and that there is *not exactly one map* that makes the diagram commute. Let us further explain the idea of the product with a miniature example of pattern recognition (see Fig. 3.6):

Fig. 3.6 The product as a pattern

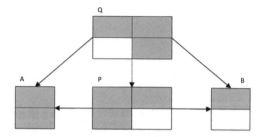

Let P be a pattern and let A and B be two factors produced from it. If pattern Q can be decomposed into the *same* factors, then it means that there is *exactly one* map from Q to P and that Q and P are *isomorphic*. Trivial as it may sound, this small example shows how the abstract universal structure of the product can be used to establish isomorphism and hence an object's invariance. That is, the idea that Q may be considered as the same object as P may be established through the universal structure of the product and the idea of projection to lower-level factors.

The Co-product

The complement of the product is the *co-product*. To understand this notion, we must first consider the idea of *duality*. The *dual notion* of a category C is constructed by replacing the domain with the co-domain, the co-domain with the domain, and $h = g \circ f$ with $h = f \circ g$ (Goldblatt, 1979, p. 45). In other words, we simply reverse the arrows in our category. The *co-product* is the dual notion of the product, generated by reversing the arrows in the category. The co-product, or *sum* of objects, is defined as follows: a co-product of C-objects A and B is a C-object A + B together with a pair $(i_A : A \rightarrow A+B, i_B : B \rightarrow A+B)$ of C-arrows such that for any pair of C-arrows of the form $f: A \rightarrow C, g: B \rightarrow C$, there is exactly one arrow $[f, g]: A+B \rightarrow C$ that makes the diagram in Fig. 3.7 commute in such a way that $[f, g]^{\circ} \, i_A = f$ and $[f, g]^{\circ} \, i_B = g$. $[f, g]$ are called the co-product arrow of f and g with respect to injections i_A and i_B (Goldblatt, 1979, p. 54).

Fig. 3.7 The co-product

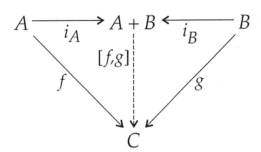

Like the product, the co-product is a way of seeing the general object ("the best of its type") from the perspective of the *particular* objects. In this sense, it is the *least specific* object to which the objects admit morphisms. This difference between the "most general" and the "least specific" may seem trivial because we usually and unconsciously conceive the particular in light of the general and the general in light of the particular. For example, the concept of Fruit is constituted through our acquaintance with particulars (e.g., Apple, Cherry, Orange), while at the same time it recursively allows us to identify the particulars as instances of the general concept. This is an interesting idea if it is framed within the discussion of reversibility and irreversibility and when it is considered in relation to the nature of wholeness as a *recursive* process in which the parts and the whole mutually constitute each other.

I previously introduced the idea that irreversibility is inevitable in natural computations as some information must be lost in order for macro-level structures to be produced. Computing the category of Fruit is therefore an irreversible process. However, when encountering a new Apple, we may easily tag it as a Fruit, which shows that the "best of its type" is probably the one that *optimally balances* the loss of information when moving upward and the gain in information when examining objects lower in the mental taxonomy.

We may illustrate the meaning of the product and the co-product in the area of natural language too. However, to address this challenge, we should understand how meaning may be represented through the idea of a *context vector*. In many practical areas of language computation, it may be helpful to represent the meaning of a word by using the words collocated with it in a large linguistic corpus. That is, we search the corpus for the appearances of our target word and examine which words appear with it in a predefined lexical window.

For instance, let's assume that we would like to understand the meaning of the word Hotdog. By searching the Corpus of Contemporary

American English (Davies, 2009) for the words collocated up to four positions to the right and left of the word Hotdog, we identify collocations such as Bun, Ketchup, and Mustard. We may use these words as a *basis* for a context vector, in which the values are the collocations' frequencies or probabilities. This context vector may be used to represent the meaning of Hotdog. Given the represented meaning of Hotdog as a context vector, we may ask questions such as how effectively can we map the meaning of Sausage to the meaning of its hyponym Hotdog as follows:

$$\text{Sausage} \rightarrow \text{Hotdog}$$

To answer this question, we first represent the meaning of Sausage as a context vector by applying the same procedure we have used before. We may now form a *unified* semantic space of Hotdog and Sausage by building a new basis that is composed of the union of the unique words that appear in the context vectors of Hotdog *and* Sausage. This new context vector may include the following words: Hot, Italian, Spicy, Potato, and Pork. The specific values of these context vectors are presented in Table 3.1 and represent the normalized percentage at which each of the basis words appears in the context of Sausage or Hotdog:

Table 3.1 The context vectors of Sausage and Hotdog

	Hot	Italian	Spicy	Potato	Pork
Sausage	10	70	5	10	5
Hotdog	60	0	20	0	20

Now, given that we are familiar with the distribution of words that are collocated with Sausage and that define its meaning, we may ask how useful is this distribution in approximating the distribution of Hotdog's context vector. This is actually a question about the possibility of mapping the meaning of Sausage to Hotdog. To answer it, we may use the Kullback–Leibler divergence (KL) measure. This is an *asymmetric* measure of the difference between two probability distributions P and Q. This measure can be understood as a measure of information gain when one revises one's beliefs from the prior probability distribution Q to the posterior probability distribution P. A value of 0 means that the distributions are the same and no information has been gained. The higher the value of the measure, the more information we have gained when revising our beliefs from the first to the second distribution. The measure is defined as follows:

$$D_{KL}\left(P\|Q\right) = \sum p(i)\log\frac{p(i)}{q(i)}$$

When mapping the meaning of Sausage to Hotdog, we may signify the distribution of Sausage as Q, signify the distribution of Hotdog as P, and apply the KL measure. The number produced through this procedure will give us an indication of the extent to which we should revise our "understanding" when trying to approximate the meaning of Hotdog through the meaning of Sausage. Again, the KL measure is *asymmetric*, meaning that approximating the meaning of Hotdog through Sausage may give us a different result from approximating the meaning of Sausage through Hotdog.

Now let's return to the product and illustrate it through the idea of *semantic transparency*. Semantic transparency describes the extent to which the meaning of a word compound, such as Hotdog, is conceived to be transparent through the meaning of its components. For example, the word compound Stomachache may be considered as a relatively simple "sum" of Stomach and Ache. We may simply understand Stomachache as an Ache associated with the Stomach. However, what about the semantic transparency of the word Mushroom? A Mushroom is a fungal growth, but a Mush is a soft, wet, pulpy mass, and a Room is simply a living space. It is quite a mystery how these two words have been glued together over the course of history to form the compound word Mushroom. Is it possible to measure the semantic transparency of a compound by using the idea of the product and the co-product?

Let's again take the word compound Hotdog as an example. If Hotdog is a semantic product of Hot and Dog, then the context vector of Hotdog should tell us something about the context vectors of Hot and Dog. The context vector of Hot may include some words, such as potato and sauce, that also appear in the context vector of Hotdog. However, the context vector of Dog probably includes a lot of words, such as cat, owner, and leash, that we won't find in the context of Hotdog. Therefore, the context vector of Hotdog may be informative about the context vector of Hot but not about the context vector of Dog.

Fig. 3.8 The Würstchen
and the Hotdog

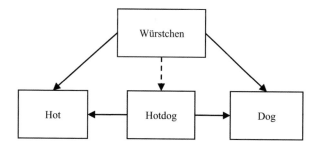

The scores of D_{KL} (Hot‖Hotdog) and D_{KL} (Dog‖Hotdog) may therefore be positively correlated with the semantic transparency score of the word compound. In this sense, breaking the meaning of Hotdog into two factors doesn't seem to be a feasible task and we may ipso facto conclude that Hotdog isn't a simple product of its word compounds.

However, a product as a universal structure is not only about projections. We can take the idea of a product a step forward. Hotdog is described in German as Würstchen. As Würstchen and Hotdog should be synonymous in meaning, we may assume that Würstchen can also be decomposed into Hot and Dog and that there should be a unique path from Würstchen to Hotdog that will make Fig. 3.8 commute.

Applying the KL divergence measure to the diagram in Fig. 3.8, we may hypothesize that the KLs of Würstchen and Hotdog should be symmetric as they are synonyms and therefore that the map of Würstchen to Hotdog should be unique, up to isomorphism, in the sense described above. That is:

$$D_{KL}\left(\text{Hotdog}\,\|\,\text{Würstchen}\right) = D_{KL}\left(\text{Würstchen}\,\|\,\text{Hotdog}\right)$$

In other words, both Hotdog and Würstchen are represented using the distribution of their context vectors. Here we assume that approximating the meaning of Hotdog through the meaning of Würstchen will yield the same D_{KL} score as the one we get when approximating the meaning of Würstchen (i.e., P in D_{KL} equation) through the meaning of Hotdog (i.e., Q in D_{KL} equation). The extent to which Hotdog is a semantic product of Hot and Dog may therefore be measured by the extent to which Würstchen is informative about Hot and Dog in comparison to the extent to which Würstchen is informative about Hot and Dog when factored through Hotdog. If Würstchen and Hotdog are perfect synonyms and their context vectors are exactly the same in German and in English, then the KL score of

$$D_{KL}\left(\text{Hotdog}\,\|\,\text{Würstchen}\right) = D_{KL}\left(\text{Würstchen}\,\|\,\text{Hotdog}\right) = 0$$

and the energy invested in approximating Hot through Würstchen and Hot (or Dog) through Hotdog and Würstchen should be the same:

$$D_{KL}\left(\text{Hot}\,\|\,\text{Würstchen}\right) = D_{KL}\left(\text{Hot}\,\|\,\text{Hotdog}\right)$$

and

$$D_{KL}\left(\text{Dog}\,\|\,\text{Würstchen}\right) = D_{KL}\left(\text{Dog}\,\|\,\text{Hotdog}\right)$$

Summary

- Category theory is a mathematical field dealing with categories, where a category is a structure composed of objects and relations (morphisms).
- Surprisingly, a category may be defined in terms where relations have precedence over objects.
- This chapter introduced several terms, such as *initial object, isomorphism, product,* and *co-product.*
- This chapter showed how the structures of the product or co-product can be used to model patterns.
- Interestingly, the product is a structure that identifies an object that is the *best of its type* while the co-product identifies the *least specific object.*

Chapter 4
How to Trick the Demon of Entropy

The previous chapter introduced category theory in order to better model the notion of structure and to address the problem that Piaget faced when he tried to model structure. Our next aim is to "trick" the demon of irreversibility (i.e., entropy) and to formulate structure in the context of irreversible processes of computation. Our first step is to introduce *the equivalence of identity and inverse in category theory* in order to better model structure.

With regard to multiplication (product), the terminal object functions as the identity object in the sense that, given object B and a terminal object **1**, $B \times \mathbf{1} = B$ (Lawvere & Schanuel, 2000). This idea can be explained as follows. To prove that object B is a product of B and **1**, we must have two maps (pr_B: B $\rightarrow$ B and pr_1: B $\rightarrow$ **1**) that satisfy the property of product projections (see Fig. 4.1).

There is only one choice for pr_1 (as it leads to the terminal object), and the map pr_B is an identity map as it leads from B to itself. Given object C with projections f to B and g to 1, there is exactly one map from C to B, which is f! See Fig. 4.2.

This line of reasoning proves that *the terminal object functions as the identity object for the product* and by the duality principle that *the initial object functions as the identity object for the co-product*. We are now in a situation where we can move forward and identify the inverse object in categories.

Under the heading "Can objects have negatives?" Lawvere and Schanuel (2000, p. 287) insightfully suggest that, if A is an object of a category, a "negative" of A means an object B such that $A + B = 0$,

© Springer International Publishing AG 2017
Y. Neuman, *Mathematical Structures of Natural Intelligence*, Mathematics in Mind,
https://doi.org/10.1007/978-3-319-68246-4_4

Fig. 4.1 The product's
projections

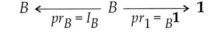

Fig. 4.2 Product and
identity

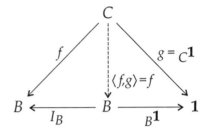

Fig. 4.3 The co-product of
sets

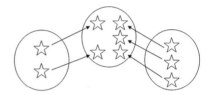

where + represents the co-product of objects, = represents isomorphism, and 0 is the initial object. For example, in the category of sets, the co-product of A and B is the set composed of the elements of the two sets (see Fig. 4.3).

It is clear that, if the co-product of A and B is zero (i.e., empty set), then *both* of these sets have to be empty sets, because in the category of sets, the empty set is the initial object. We can ask how general is this idea and whether A and B must be initial objects if A + B = **0**. The proof is as in Fig. 4.4:

Fig. 4.4 The proof

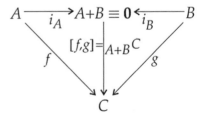

Here, *f* and *g* are two morphisms that are the same as [*f*, *g*], as a map is defined by what it accomplishes and in both cases we end up with C. There is only one map [*f*, *g*] as its domain is an initial object. Therefore, there is only one map from B to C, which means that B is an initial object and the same is true for A.

Fig. 4.5 The groupoid

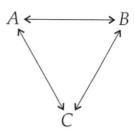

The striking conclusion is that, with regard to the co-product, *only initial objects have negatives* and their negatives/inverses are *themselves initial objects*. That is, if A + B = 0, then A = B = 0. This conclusion, which is fully explained by Lawvere and Schanuel (2000), holds for the dual – the product – where only terminal objects have negatives that are terminals themselves. That is, if A × B = 1, then A = B = 1.

As I've argued (Neuman, 2013), this conclusion is far from trivial. If only initial objects in a category have negatives, *then reversibility is guaranteed only for them and only under the condition of the co-product*. As I've further argued, this statement means that reversibility through the inverse can exist only in a limited kind of structure. Interestingly, if with regard to the product only terminals have negatives, then a terminal object has a negative only in the company of other terminals. See Fig. 4.5 for an example.

The triad in Fig. 4.5 is composed of isomorphic terminal/initial objects that mutually constitute equivalence relations, as each of them factors through the others. They are "isomorphic sub-objects" of each other (Goldblatt, 1979). These are objects that are isomorphic and whose relations among themselves, here described in terms of information flow, are *reversible*. This structure ensures reversibility and is known as a *groupoid*; it is a small category in which *each morphism is an isomorphism*, or "a category in which each edge (morphism) is invertible" (Higgins, 2005, p. 5).

Based on the above line of reasoning, I have argued (Neuman, 2013) that the groupoid is the theoretical *building block* for modeling a structure in cognitive systems and hence that the groupoid is the building block of structures in natural intelligence. At this point, important qualifications should be added. I don't identify the groupoid with a structure. I only argue that it is a building block for understanding structure. It is not the only building block. Moreover, I don't pretend to identify the mathematical holy grail of natural intelligence. The groupoid is a building block, but as a mathematical structure, it is too limiting, and therefore its sense should be loosened and contextualized before it can be used as a productive component in modeling structures.

So far, we have come some way, moving from Piaget's structuralist thesis to the idea that the groupoid is a building block of structures in natural intelligence. At this point it is vital to consider the groupoid and the notion of symmetry in more "dynamic" terms and to examine the way in which it may resolve the irreversibility quandary.

We have considered isomorphism in "structural" terms, but we may also think about it in terms of *synchronization*. To explain this point, we should understand the deep connections between the groupoid and synchronization, as *synchronization may be the important equivalent of reversibility/isomorphism*. Golubitsky and Stewart (2006) show that the symmetry formed by the groupoid implies *synchrony and similar periodic dynamics*. For example, the following network is composed of two cells, called x_1 and x_2 (Fig. 4.6):

Fig. 4.6 A two-cell network

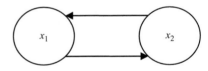

This network, which is a groupoid, forms a cyclic group of order 2. This system can be described through the following equations:

$$x_1 \# = g\left(x_1, x_2\right)$$
$$x_2 \# = g\left(x_2, x_1\right)$$

and, for period T, $x_2(t) = x_1(t + T/2)$ for all t, which means that the state of cell x_2 at time t is a function of x_1 at time t + half a period.

In sum, interpreting the isomorphism of the groupoid in dynamical terms is interpreting reversibility as some kind of synchronization or more accurately as *dynamic coupling* between interacting units. In this way, reversibility may exist in limited portions of our cognitive system but only temporarily before being sacrificed for the benefit of higher-order structures, an irreversible process that necessarily involves the loss of some information. This interpretation of the groupoid will be used in the next chapter, which includes a demonstration of the relevance of the groupoid for understanding neural network connectivity.

Summary

- In this chapter, the task was to find the equivalence between "identity" and "inverse" in category theory.
- The terminal object functions as the identity object for the product, and the initial object functions as the identity object for the co-product.
- With regard to the co-product and product (respectively), only initial and terminal objects have "negatives."
- Reversibility is guaranteed for the product and co-product through these "negative" objects.
- The groupoid is a structure in which reversibility is guaranteed; it may therefore be used to model the building blocks of a structure.

Chapter 5
Neural Networks and Groupoids

The chapters so far have shown that the groupoid may be considered to be a building block of a structure and that the groupoid may have a dynamic nature. In this chapter, I explain and illustrate this idea by examining it in the context of neural networks. Exploration of the unique computations performed by the human brain, as well as developing brain-inspired machine learning algorithms, should involve a deep understanding of neuronal circuits and their functionality. In this context, the architecture of cortical connectivity in the mammalian brain is usually discussed in terms of a *hierarchy*, where information mostly flows in either a bottom-up (i.e., feed-forward) or top-down manner between a sequence of cortical layers (e.g., Grill-Spector & Malach, 2004). For instance, the HMAX computational model of the visual cortex describes vision in terms of the hierarchical flow of information from the primary visual area upward (Poggio & Serre, 2013).

In the field of brain-inspired machine learning algorithms, the focused interest in the hierarchical architecture of neural networks may be partially attributed to popular models inspired by (but not necessarily rigorously adhering to) cortical computations. One example is the enormously popular "deep learning" model, which is an approach that basically involves a bottom-up process of computation. The reliance of the deep learning approach on hierarchical architecture doesn't contradict the possibility of using recurrent circuits within this framework, as demonstrated in the idea of recurrent neural networks (e.g., long short-term memory), where connections between units are cyclical and "horizontal."

© Springer International Publishing AG 2017
Y. Neuman, *Mathematical Structures of Natural Intelligence*, Mathematics in Mind,
https://doi.org/10.1007/978-3-319-68246-4_5

The existence of a cortical hierarchy is also compatible with the experimental findings, which indicate that most cortical connections are local, recurrent, and excitatory (Amir, Harel, & Malach, 1993; Douglas & Martin, 2007; Moutard, Dehaene, & Malach, 2005) and that this architecture of the neocortical circuits plays a crucial role in natural computation, which generally outperforms human-made systems in its efficiency. In other words, "horizontal connections" (Amir et al., 1993) that form overlapping clusters (Malach, Harel, Amir, & Grinvald, 1993), described by Sporns and Betzel (2016) as "network modules," seem to play a crucial role in cortical computations.

It is suggested that these modules, which involve dense and reciprocal short-range connectivity, produce *local integration of information* (Fisch et al., 2009; Malach, 2012), which means that information is integrated by these modules *before being propagated* through long-range connections to higher levels of the network assembly. The exact meaning of this dynamic is far from clear, and explaining it is a challenge I attempt to surmount in this chapter.

Our point of departure is the idea that the mammalian brain, as one specific instance of natural intelligence, is basically involved in the identification of structures/patterns. In the brain, information must be adequately coordinated to produce patterns to be memorized and used for various cognitive tasks, from object recognition to planning. These patterns are what have been described by some of the old-school psychologists and neuroscientists as "structures" or "Gestalts."

It must be noted that the term "structure" may be used in two discrete epistemological senses. The first sense involves a description of a *concrete* perception as experienced by a biological–cognitive agent such as a human being. This is the experience from "within" or what has been described as the phenomenology of the subject, for instance, when it recognizes a certain object (e.g., a car). The second sense involves the *abstraction* and *formalization* of the concept of structure. In other words, the first sense involves the experience from within, the first-person perspective, while the second sense involves the study of the abstract concept of structure. In this chapter, I adhere to the second sense, which I have already explained in the preceding chapters.

The existence of structures/patterns in the brain seems to offer an optimal solution for addressing the capacity limit facing natural information processing. It has been proposed that the nervous system relies on at least two mechanisms that counteract its capacity limit in processing information: *compression* and *forgetting* (Tetzlaff, Kolodziejski, Markelic, & Wörgötter, 2012). That is, and similarly to information compression in human-made systems (Salomon, 2007), it is argued that the nervous system compresses information in some cases and throws it out in others in order to avoid overload.

In this way, and as may be explained in terms of information theory (Timme, Alford, Flecker, & Beggs, 2014), the formation of a structure, such as in the case of forming semantic categories (Tamir & Neuman, 2015), is a practical solution to the problem of capacity limits, as the structure provides the brain with an optimal combination of *forgetting* (i.e., throwing away some information about the specific instances that form the structure) and *redundancy* (i.e., information concerning regularity; Grunwald, 2004) for information compression. This point was mentioned in our discussion of the product and the co-product (see Chap. 3), and it may be illustrated as follows.

In the process of forming a representation of a cat, some information concerning particular cats we have encountered throughout our life is thrown away, while other information that is redundant is used to compress the information into a more economical representation of the cat for future use.

Moreover, this structure is used to build novel structures, such as the metaphor "cat," which denotes a jazz player. Without forgetting some information about real cats and compressing other information, we could not form the template/structure of an actual cat. Similarly, without the abstract template of an actual cat, the metaphor of a cat couldn't be formed either. Structures, as wholes, are multilayered and composed of various levels of organization that include wholes and sub-wholes.

From this standpoint, and given the argument that the groupoid is a building block of natural intelligence, we should ask whether we may use the groupoid as a model of local recurrent cortical circuits. These network modules are functionally described as *local integration units* (Fisch et al., 2009; Malach, 2007). As such, they must constitute some form of assembly that can be guaranteed to have the quality of reversibility and hence that has the unique structure of the groupoid. Moreover, the "recurrent" tag of these modules means that information is fed back and forth between their constituting units. This recurrence seems to correspond with the idea of the groupoid as a structure in which every morphism is an isomorphism.

To explain this point, we should recall the deep connections between the groupoid and synchronization, *as synchronization may be the important equivalent of reversibility/isomorphism in cortical circuits.* If we conceive the groupoid as a model of recurrent and local circuits, then we may understand that one of its benefits may be in modeling the synchronization of neurons. However, a groupoid structure has more benefits than simply in expressing the orchestrated activity of cells in response to a stimulus (i.e., a real-world pattern), and this thesis is further elaborated here.

I have argued that the groupoid is an abstract structure that may be used to model a structure in general and cortical local recurrent circuits in particular. The main argument may be repeated as follows. A structure, at least according to Piaget, requires reversibility and hence the existence of the inverse function. How can we model the inverse in abstract systems where objects/nodes are interconnected? We used category theory to address this challenge and identified the groupoid as the structure that supports the inverse function. From Golubitsky and Stewart (2006), we learned that the groupoid is characterized by isomorphic relations that entail synchronization/coupling and can conclude that synchronization is the equivalent to what Piaget described as reversibility; the emerging property of the groupoid, which is a structure constituted by the inverse, is synchronization.

The groupoid's isomorphic (i.e., reversible) relations are important not only from a theoretical perspective, as has been discussed so far, but also (and mainly) from a physical computational perspective. To recall, the main insight of the physics of computation, epitomized by *Landauer's principle* (Bennett & Landauer, 1985), is that there is a well-defined energetic price a system must pay during irreversible computation – that is, a process in which the inputs cannot be fully reconstructed given the output and the binary function operated on the inputs. As the brain is a physical computational system, it may be helpful to model natural computations using the idea of information loss in an irreversible computation.

For example, the hierarchical architecture of the brain entails the loss of information when higher-order percepts are formed from lower-order representations. When we form the percept of a cat, for instance, and store it in our memory, we don't preserve all the perceptual information we have gathered during our encounter with each and every cat we have seen. As I have outlined, our brain actually throws out a huge amount of information about these concrete cats in order to build a more abstract and informationally economical representation that will serve it in the future. The economical aspect of this representation may be evident in the relatively small number of instances the human brain requires in order to represent a concept in comparison with the greater number of instances per class required by artificial systems, as evident in models of deep learning, for instance, which according to the leading figures in the field require 5,000 cases per class to gain good classificatory performance (Goodfellow, Bengio, & Curville, 2016).

It follows, then, that the "network module" formed out of "horizontal" and reciprocal short-range connectivity may function as an information-preserving patch of neurons that, through their synchronized/coupled activity, as implied by the groupoid topology, *integrate the information flowing upward before some of it is lost.* By "integrate" I mean that by being synchronized, the local and recurrent network may function as a *cluster in which some unique information produced by the neurons is selected out in favor of a macro-level state, which in turn is propagated upward as a whole signal.*

In other words, the coupled activity of neurons, through local and recurrent circuits, forms a *temporary* assembly that functions as a cluster of information (see Malach, 2012). It therefore seems that the groupoid topology may be used to naturally compress information and as such to overcome the combinatorial complexity that accompanies the multilayered representations of the brain and artificial systems alike.

To explain and illustrate this point further, we may use an oversimplified and clearly wrong *model* of vision. The brain may be considered as a camera that represents a stimulus (e.g., cat) as an array of pixels. As such, it is like an intelligent camera that breaks the represented stimuli (e.g., a percept of a cat) into a basic set of partitions of pixels and forms the general pattern (e.g., of a cat) in a bottom-up manner from these basic components. In other words, we may (wrongly) model the brain as representing the cat in an array of independent pixels and building the template of a cat from increasing levels of these pixels' combinations (two by two pixels, four by four pixels, and so on). This idea is clearly in line with the architecture of deep neural networks and specifically the theory of hierarchical temporal memory (George, 2008; George & Hawkins, 2009), in which the architecture consists of multiple layers of representations.

The problem is that the number of potential partitions of an n-element set (e.g., the set of pixels) is the Bell number, which means that for a 10-element set (e.g., pixels), we have 115,975 possible partitions (i.e., potential breakdowns of the set); for an 11-element set, we have 678,570 potential set partitions; and so on. Given the finite capacity limits of the brain, such an idea is improbable unless we can *force some constraints* on how representations are formed from basic representations (for some solutions, see George, 2008).

Therefore, the metaphor of the brain as an intelligent camera is clearly wrong as the combinatorial complexity of the particles makes no sense without the existence of an assembly that is formed through constraints imposed on the representations. However, things may look different if we use the groupoid structure to model local and recurrent circuits, as the groupoid may force the required constraints on lower-level representations. To further explain the function of local recurrent circuits, we may want to adopt a better metaphor for an intelligent system – specifically, natural language (see Malach, 2012).

Human written language is made up of a small set of letters. These letters may be organized in specific combinations to form units of meaning – words – and these units of meaning may be organized further to form higher-order structures of meaning (i.e., propositions, sentences, etc.). On the one hand, the formation of words *reduces* the combinatorial complexity entailed by the potential combination of letters, and on the other hand, it allows the *increase* of the combinatorial complexity of higher levels by allowing us to form a potentially infinite number of sentences. That is, the constraints imposed by a bottom-up construction are not only a way of reducing complexity at a lower level of analysis but also a way of building the potential for *increasing complexity* at higher levels of analysis. Increasing complexity is highly important as it is through this expanding field of possibilities that creativity, novelty, and the potential for new solutions become possible. This is a central point[1] that should be repeatedly

---

[1] I'm grateful to Rafi Malach for raising this point.

emphasized as the discourse concerning artificial intelligence systems is usually occupied with reducing the combinatorial complexity of well-defined tasks. Natural intelligence doesn't work this way, as it involves both a decrease and an increase in complexity and for very good reasons. With this insight in mind, we may return to the groupoid.

If we add to the abovementioned hierarchical architecture local, lateral, and recurrent connections that form groupoids, then micro-structures (i.e., assemblies) are formed in a way that involves "local integration," which reduces the potential number of higher-order structures given the combinatorial explosion of a bottom-up construction from basic features.

The combinatorial complexity of the brain *potentially* corresponds with the huge combinatorial space of neural activity. I use the term "potentially" as actually the neurons in the brain are not fully connected. Even if we take this constraint into account, the potential activation space of the connected neurons is still huge. When some of this neural activity is synchronized through recurrent and local short-range connectivity, this overall combinatorial complexity may be significantly reduced, as some assemblies of the neural activity will turn out to be less probable than others.

On the other hand, the formation of these assemblies, or units of meaning, may allow the formation of "micro-spaces" within which the combinatorial complexity of each assembly will greatly increase the complexity of the templates we use in order to function in the world; in other words, to recognize a human face, we don't have to represent all of its possible combinations (e.g., with or without a moustache).

To push the language metaphor a step forward, we can say the same thing using different words just as we can use the same word to say different things, in both cases, using a small, finite set of letters, a huge combinatorial space, forced constraints, and multiple levels of construction. This flexibility, which is evident in various intelligent systems, such as the immune system, may be modeled through increasingly high levels of assembly integration.

The main argument presented so far may be further explained from the perspective of information theory, through the notion of *information decomposition* (Williams & Beer, 2010). For simplicity, this idea will be illustrated with a network of three units. Williams and Beer (2010) suggest that two variables' total information about a third variable (i.e., $I(X_1; X_2, Y)$) can be decomposed into four nonnegative terms:

$$I\left(X_1, X_2; Y\right) \equiv \mathrm{Synergy}\left(Y; X_1, X_2\right) + \mathrm{Unique}\left(Y, X_1\right)$$
$$+ \mathrm{Unique}\left(Y, X_2\right) + \mathrm{Redundancy}\left(Y; X_1, X_2\right)$$

In other words, the total information of the three units, which in our case correspond to neurons, can be decomposed into the sum of the unique information provided by each "predictor" (i.e., X_1 and X_2) about the output variable Y, the redundant information provided by both predictors about Y, and the synergy that is the information provided beyond the unique information provided by each predictor and the redundant information provided by both of them.

We may better understand the role of the groupoid and local recurrent networks in terms of information decomposition. Again, for simplicity, we will stick to the case of the three-unit groupoid. The unique topology of the groupoid can be interpreted as a structure in which the synchronization implies that the information redundancy provided by any two units with regard to the third is always higher (beyond a certain threshold) than the unique information provided by each of them. If the redundancy of a given set of units is significantly higher than their unique contribution, as detailed above, then we may "forget" the information *uniquely contributed by each unit* and consider the whole assemble as a cluster that may be propagated upward for further processing.

This point may be further explained formally and with regard to a three-unit groupoid (i.e., a local recurrent network) and a small example algorithm. Let G be a set of N units (e.g., G_1, G_2, G_3). We then compute the mutual information of each combination by choosing $N-1$ units from the set. In the case of three units, the number of produced combinations is three, as follows:

$$I_i\left(G_1; G_2, G_3\right)$$
$$I_i\left(G_2; G_1, G_3\right)$$
$$I_i\left(G_3; G_1, G_2\right)$$

where i is a running index from 1 to N. For $i = 1$ to N, we next compute the unique information provided by each unit with regard to the output as well as the redundant information produced by each of two units with regard to the third. If the redundant information across all combinations exceeds the unique information, according to some criteria, then we merge G_1, G_2, and G_3 into a single new assembly, G, and propagate it to the next layer of cortical computation. This simple algorithm may explain how local integration may be formed through the structure of the groupoid and how it may reduce combinatorial complexity.

It was argued by Phillips, Clark, and Silverstein (2015, p. 2) that local cortical circuits "synchronize selected signals into coherent sub-sets and therefore form dynamic grouping by synchronization" (p. 3). My colleague Rafi Malach (2012) has proposed that, in such a structure, the information about the state of a local neuronal assembly is distributed back to the neurons that form the assembly through recurrent activations. As the groupoid structure implies synchronization, it is clear why it is a relevant structure for modeling such an activity.

Modeling cortical circuits through the groupoid doesn't mean that the neurons are directly connected but just that we are dealing with a network whose components are physically close (Kopell, Gritton, Whittington, & Kramer, 2014) and synchronized – that is, the activity generated by each member of the assembly is fed back to the original member within a relatively short period of time (typically less than a second). When we model local recurrent circuits through the groupoid, we therefore assume that synchronization is taking place locally (i.e., in topographically limited regions of the cortex), an assumption that in its turn implies that the computations are faster in comparison with long-range connections.

The differences in *distance and time scales* between short- and long-range connections are highly important as it allows the information flowing upward to be *momentarily and locally integrated before some of it is lost*. It means that the local recurrent circuits form information clusters before any of the information flowing upward becomes subject to forgetting or compression. As explained by Sporns and Betzel (2016, p. 615), a network module is a "sub-network of densely interconnected nodes that is connected sparsely to the rest of the network." Given a locally dense, but a functionally isolated, subnetwork of neurons that are synchronized through mutual activation and amplification, it is possible (as described above) to form – through what we might term the *redundancy over uniqueness* heuristic – "patches" of integrated information that can be propagated forward to further cortical computations (see also Harmelech & Malach, 2013).

At this point, we may speculate that designing brain-inspired algorithms using the abovementioned modeling of recurrent local circuits may show some promise for the field, specifically for researchers developing networks along the lines of the information theory ideas presented above. The thesis presented in this chapter may explain how a certain abstract structure (i.e., the groupoid), which theoretically has been found to provide the basis for reversibility, the inverse function, and structure formation, is theoretically relevant for explaining real and "natural" neural information processing. I have proposed that the idea of the groupoid is epitomized by the synchronized activity of neurons that are physically close and therefore co-activated as a response to the appearance of associated stimuli. Instantiating the meaning of the groupoid in the context of neural networks was a temporary shift in the narrative of this book, used only to justify the explanatory power of the groupoid as a building block for the formation of structures. Therefore, we return to deal with the abstract aspects of structure in the next chapter.

Summary

- Neural circuits in the mammalian cortex are usually discussed in terms of a hierarchical structure.
- Most cortical connections are local, recurrent, and excitatory.
- The groupoid may model this architecture.
- The groupoid structure of these neural modules is composed of coupled neural activity.
- The groupoid structure integrates information flowing upward before some of it is lost as a result of an irreversible process of computation.

Part II

Chapter 6
Natural Intelligence in the Wild

In the first part of this book, I presented the outlines for a neo-structuralist agenda that strives to identify structures for modeling natural intelligence; in this effort, I was both reflective on and critical of past ventures. More specifically, I focused on Piaget's notion of structuralism and addressed the reversibility/irreversibility issue that bothered him. I also proposed the idea that the groupoid may be used as a building block of structure. I justified this idea theoretically based on an attempt to formalize structure along Piagetian lines, and I illustrated it in the context of the mammalian neural network. I see the notion of the groupoid as expressing *local symmetries* that may be the building blocks of a structure; however, I have no fixation on this concept as a magic bullet for resolving the whole quandary of natural intelligence.

In this part of the book, I add layers of complexity to the basic analysis presented in the first part and attempt to enrich our understanding of the way structures exist in vivo, outside the laboratory of clean mathematical formalism. This attempt will require careful adaptations, hypotheses, and speculations, all of which seem to be inevitable when walking on the wild side of life.

As I've emphasized before, natural intelligence (which lives "in the wild") may not easily give itself over to abstract and mathematical theorization. Therefore, we must clarify the meaning of groupoid in this wild context and if needed to loosen it so it better fits as an explanatory concept.

© Springer International Publishing AG 2017
Y. Neuman, *Mathematical Structures of Natural Intelligence*, Mathematics in Mind,
https://doi.org/10.1007/978-3-319-68246-4_6

For reasons of didactical exposition, we will assume that objects such as flowers, bees, lions, and apples exist and may be formalized as categories. Later I will try to explain how the structure of such objects emerged in the first place, but, for the current phase, let me start with the commonsense observation that when organisms interact with objects they usually interact with them *along a time line*. That is – and in contrast with some artificial intelligence systems of pattern recognition, such as those of deep learning – natural intelligence is not exposed to isolated and artificial pictures of various objects instantiating a given set.

In nature, a wolf is not exposed to 5000 pictures of rabbits; rather, it tracks a single living and moving rabbit along time, in a temporal and spatial context, and by experiencing the object in all modalities (e.g., smell and sound) and as deeply woven into the social context of the wolf's community. It is therefore probable that natural intelligence works along different lines from those of most machine learning procedures. The wolf doesn't have in his mind "training" and "test" sets of rabbits, although one may argue that this is precisely the way evolution has shaped the wolf's mind.

Now, tracing an object through time, as one instance of natural intelligence, may be motivated by the epistemological axiom that *successive appearances of a given object along time are isomorphic* (or at least similar up to isomorphism). This epistemological axiom is actually grounded in the ancient notion of *association*, as introduced by Aristotle, John Locke, David Hume, and others. It means that multiple instantiations of an object, whether mental or concrete, that repeatedly appears in time and space may be conceived to be associated and somehow similar regardless of any surface variations.

The appearance of an object along a limited context of time and space involves some kind of continuity. *Natura non facit saltus* – "nature does not make jumps" – and therefore the appearance of the same object with no "jumps" implies that it has a single identity. Natural things change gradually unless they are subjected to a catastrophic event (in this case, they are considered to be violations of an expected order, which surprises us). Therefore, in most cases we are able to learn about objects over time, through minor and gradual variations. Natural objects are presented to us at spatially and temporally close intervals, and therefore, when we track a single object through

Fig. 6.1 An apple in transformation

Fig. 6.2 The first three
phases of consumption

Fig. 6.3 The consumption
as a groupoid

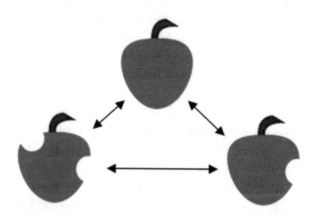

time, we may consider the gradual transformation of an object and
assume that its various appearances are *isomorphic*, at least as a first
working hypothesis. Let's observe Fig. 6.1, in which an apple has
been traced through its consumption process bite by bite.

You can see that the final object (to the right) is quite different from
the first object (to the left). In fact, you may even wonder whether
there is a meaningful *structure-preserving transformation* relating the
first apple to the eaten apple. Let's examine the first three steps of the
process under the association assumption (see Fig. 6.2).

As the mapping from the first object to the second and third is *by
assumption* an isomorphism, we may represent the first three steps as
a groupoid (see Fig. 6.3).

This is a groupoid composed of three objects forming the equivalent class of apple. We may construct groupoids of four objects, five

objects, and so on. However, we should acknowledge that, when exposed to natural objects in natural situations, we are exposed to limited and bounded *episodes*. This means that the size of the formed groupoid may be limited in practice. According to this logic, natural episodes are those that define the boundaries of the processed situation and therefore the nature of the emergent structure.

Natural episodes, in contrast with artificial stimuli, present us and other organisms with objects, actions, and properties that are interwoven with each other in a way that forms a dynamic pattern. This process, which is limited by our memory, may explain why natural intelligence is different from artificial intelligence and why natural intelligence is different across species.

First, different species are exposed to different episodes and therefore have different worldviews (sometimes called *Umwelten*). In other words, different species address different real-world challenges and therefore experience different episodes. The challenges faced by guppy fish are quite different from those faced by humans or parrots. In addition, *working memory* may play a key role in the formation of structures as it defines the size of the "window" through which we observe objects and therefore the size of the groupoid and the complexity of the emerging structure.

Assuming that successive objects form a groupoid doesn't yet explain how a structure is formed. On the first level, we may build a list of *structure-preserving transformations* between successive phases of the object and consider these transformations as those defining the object (e.g., apple). It is important to emphasize the idea that the structure-preserving transformations don't assume the a priori existence of a structure. They are structure preserving on the basis of the hypothesis regarding the object's successive appearances in time and as such are derived from the association between the successive objects. That is, the transformations between successive objects are hypothesized to be structure preserving and recursively "define" the identity of the object.

This is a vital point. While from a mathematical perspective structure-preserving transformations must be proved, from the epistemological perspective, the idea that the successive objects express the appearance of the *same* object (which is undergoing structure-preserving transformations) may be considered a *hypothesis* formed by our brain/mind. It is an *abductive* form of reasoning (as described by Charles Sanders Peirce). Natural intelligence seems to primarily consist of the generation of hypotheses, in contrast to machines' deductive or inductive reasoning.

At this point, we should also acknowledge the fact that, when we use the central term "isomorphism," we should doubt whether "sameness" in real-life situations can be described using such strict mathematical criteria. Therefore, identity and isomorphism may be the ultimate ideals of mathematical similarity, but in the real world, more complex and "softer" ideas of sameness must be adopted.

There is another layer of complexity that I would like to add. While, up to now, we have defined the object through a set of transformations, at a higher level of abstraction we may analyze the mapping functions between these structure-preserving transformations. See Fig. 6.4:

Fig. 6.4 A functor
between the
transformations functors

In Fig. 6.4, you can see that we've added a map from the map associating the first and second objects to the map associating the second and third objects. This added map is a *functor* that exists at a higher level of abstraction than functors we have encountered before (see Chap. 3). The map between the first appearance of the object and the second appearance of the object is a functor, as is the map between the second appearance of the object and the third appearance of the object. The higher-level functor is *a functor between these functors*!

If natural intelligence exists along the lines described up to this point, then we have another explanation for why, and only in certain contexts, human beings may form more complex structures than other non-human organisms. This is of course not a real "why" explanation but a descriptive explanation suggesting that what characterizes human intelligence is the ability to build *maps of maps* (i.e., meta-maps), which in turn give rise to more complex and abstract structures, as implied by Piaget's idea of reflective abstraction. In other words, the "how" of the way functors are transformed is what determines the "what" that is formed. The idea of building maps of maps (or analogies between analogies) is important and will be further explained in the following chapters.

Summary

- Time is a crucial dimension for experiencing objects "in the wild."
- Successive appearances of an object in time may be considered *by hypothesis* as appearances of isomorphic objects.
- Both transformations of an object in time and their higher-order mappings form patterns.
- Piaget's notion of reflective abstraction is evident in this context as we learn by abstracting transformations.
- However, the strong notions of identity and similarity must be set aside to account for the activity of natural intelligence in the wild.

Chapter 7
Natural Intelligence Is About Meaning

I have repeatedly emphasized the idea that natural intelligence is "relational" and that this relational architecture may be represented and studied through morphisms, giving birth to what we conceive as abstract structures. An insightful experiment conducted by the Gestalt psychologist Wolfgang Köhler may help us to better support and elaborate this idea.

The experiment (described in Luria, 1976) involved a single subject – a hen – and a very simple apparatus that involved grains and two sheets of paper. The experimental procedure was as follows. The hen was presented with grains on the two sheets of paper, one *light gray* and the other *dark gray*. On the light-gray sheet, the grains simply rested on the surface of the paper, so that the hen could peck at them and eat them, whereas those on the dark-gray sheet of paper were glued in place so that the hen could not peck at them. That is, the light-gray sheet provided the hen with a positive reward, in contrast with the dark-gray sheet.

During the learning phase, the hen was exposed to the sheets in several trials. She quickly learned the logic of the experiment, pecking at the light-gray sheet and avoiding the dark-gray sheet. At this point, the experiment moved to the next and more challenging test phase.

During the test phase, the hen was presented with a new pair of sheets, one of which was the *same light-gray* sheet and the other of which was a new *white sheet*. Now the interesting question was how the hen would behave in this case: to which of the sheets would she

© Springer International Publishing AG 2017

Y. Neuman, *Mathematical Structures of Natural Intelligence*, Mathematics in Mind,
https://doi.org/10.1007/978-3-319-68246-4_7

positively react? If the hen were "object oriented," then she should have responded to the light-gray sheet as this object was associated in her mind with a positive reward. However, the astounding results of this single case study were that, most often, the hen approached the *new white sheet*. Köhler explained these results by proposing that the hen had been directed not to the absolute darkness or lightness of the sheet but to the *relative* lightness. In other words, what basically triggered the hen's learning was an abstract *difference* or relation. This conclusion portrays natural intelligence, as expressed by the hen, in terms of a highly abstract concept – difference. However, it is important to recall that natural intelligence is driven by *meaning*, which (as I have previously proposed) is a *value-based process of mapping*.

When exposed to the light- and dark-gray sheets during the learning phase, the hen didn't respond just to an abstract mathematical concept but learned that this relation of order (i.e., one sheet is lighter than the other) was mapped onto a two-value set involving the pleasure of reward vs. a pleasureless non-reward. At this point, the difference between the dark- and the light-gray sheets turned into a "difference that makes a difference" (Bateson, 1979/2000), which according to Gregory Bateson is the basic unit of the mind. The relation between the light-gray and the dark-gray sheets during the learning phase is a relation of *order*, which can be represented by a directed acyclic graph in which the direction of order is depicted by a directed edge:

$$\text{Dark Gray} \rightarrow \text{Light Gray}$$

This simple graph almost immediately exposes the idea that, in themselves, the light-gray and the dark-gray sheets are meaningless. They are *objects formed* via *morphism in a category*, and they are objects that gain their basic sense by being defined through the order relation "darker than" or its dual "lighter than," a relation deeply grounded in the hen's perceptual system.

If you are surprised by this counterintuitive idea of things being defined through relations, then you must understand how economical it is in terms of natural computations. Let's assume that somehow the hen encounters situations in which she must choose whether to invest her energy in pecking lighter and rewarding sheets or whether to invest her energy in pecking non-rewarding darker sheets. In vivo, the hen may have encountered an enormous variety of light and dark cases.

Various ever-changing aspects – the specific wavelengths forming the physical basis of gray, light, and dark; the uncertainty associated with different lighting conditions; and the noisy world of visual stimuli – might have turned the hen's decision-making process into a nightmare. Instead of dealing with philosophical questions such as "what is the meaning of being light/dark?," natural intelligence, as epitomized in the hen's behavior, seems to have chosen a different and a more constructive path.

The path chosen by natural intelligence is to focus on *meaningful relations*, in our case the order relation, and to consider the "objects" of this relation as secondary, in such a way that they can be substituted, as manipulated in Köhler's experiment, without tricking the mind. This *relational epistemology* is radical but deeply grounded in the logic of natural intelligence. Of course, it has enormous benefit as it means organisms don't have to deal with complex philosophical and ontological issues. In an imaginary world where hens can talk, one could have interviewed Köhler's hen and asked her how she made the decision to peck at the correct sheet given the fuzzy, gradual, and uncertain nature of being "light" or "dark." The hen would probably have answered the wise psychologist that she simply pecked at the lighter thing. That's all ….

In sum, the order relation between the light- and the dark-gray sheets has formed an abstract category that becomes meaningful when loaded with value (i.e., reward). This category may be described as a difference that makes a difference, and it is composed of (1) a category with an order relation defining two objects and (2) a mapping function from this category to a basic "reward" category with two values: 1 and 0. The idea of the value or the "reward" category deserves more attention.

We may consider the difference category as a set (D) in which the lighter object (L) is a subset. It then follows that a *characteristic function* of L is a function that maps elements of D to the two-value set "2":

$$X_L : D \to 2$$

such that those elements of D in L give the output "1" (i.e., reward) and those not in L give the output "0" (non-reward). This minimalistic machinery may explain how the lighter element is conceived as rewarding and directs the future behavior of the hen through a very short and limited learning phase. The reward category is therefore composed of two objects: reward and non-reward, signified as "1" and "0," respectively. The non-reward object, like the zero in mathematics, may be represented as the *absence* of a stimulus or reward.

We may agree to accept the idea that the 0, or the non-reward object, corresponds with some absence. In the category of sets, this absence is represented as an empty set – { } – with no elements. Interestingly, in the category of sets, the empty set functions as the initial object from which arrow(s) are launched to all other objects in the category. Therefore, the reward category may be represented as a category where a morphism is directed from the initial "empty" object to the reward object "1." The hen's learning phase may therefore be represented as follows, where the difference category is mapped onto the reward category. In this line of representation, the test phase is the mapping of the first difference category to the second difference category and the reward category (Fig. 7.1).

What we actually see is that the difference/relation between the white and the light-gray sheets has been mapped to the difference between the light-gray and the dark-gray sheets. That is, we observe a process in which there is a *similarity of differences*. As insightfully proposed by the quantum physicist David Bohm (1998), *order* appears to us as the interplay of "similarity of differences" and "difference of similarities."

The difference between the light and the dark sheets has thus been mapped during the learning phase to the difference between reward and non-reward. This is of course a kind of *reinforcement learning*, but I prefer to consider the reward category as a *value category* as value may describe various forms of preferences, such as an esthetic preference or a moral preference, which cannot be trivially reduced to

Fig. 7.1 The mapping between the differences and value categories

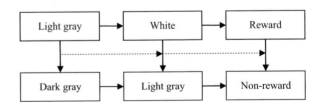

reward. In the context of the hen's experiment, I will use the term "reward" rather than the more general term "value."

I have explained the similarity of differences as a characteristic function that maps the lighter element to the rewarding experience and the darker element to the non-rewarding experience. During the test phase, the difference "white light:gray" is mapped to the difference "light gray:dark," which has already been loaded with value.

In fact, we may even argue that the similarity of differences forms an *analogy* in which light gray stands for dark just as white stands for light gray. Formally, this analogical relation is represented as follows:

$$light\_gray :: dark :: white : light\_gray$$

The mathematician Stefan Banach said that "good mathematicians see analogies" but "great mathematicians see *analogies between analogies*" (my emphasis). The hen in Köhler's experiment clearly "saw" the analogy between the sheets, although one may hardly describe the hen as a mathematician. However, her ability to identify analogies indicates that her cognition was guided by the same basic mathematical structures underlying the mathematician's cognition. From a structuralist perspective, what is so fascinating is that the same logic underlying the hen's cognition characterizes other forms of meaning-making systems.

For example, Saussure pointed out that meaning primarily involves the exchange of two elements belonging to the same domain and their relation with an element from another domain. For instance, one US dollar has value as long as it can be exchanged for a number of euros and as long as the two currencies can be exchanged for a nonmonetary element, such as a T-shirt. While there is a difference between dollars and euros, the exchange rate establishes an isomorphic map between them, and both can be exchanged for a single T-shirt. The light- and the dark-gray sheets are both elements belonging to the same domain and defined through the order relation of their lightness. However, this relation is loaded with meaning when translated into an element from another domain, which is the value category.

What have we learned from Köhler's experiment and its conceptualization? And how is this lesson related to the notion of the groupoid and the formation of structures? We have learned that there is good reason to understand that natural intelligence is deeply relational in the sense that it is grounded in certain morphisms (i.e., transformations *à la* Piaget) and that objects/elements are to be defined through these relations and not vice versa. We have also learned that natural intelligence is value based in the sense that a difference that makes a difference is always a mapping function grounded in a value system. We have also learned that similarity of differences forms higher-order structures and that complex structures may be formed through morphisms of morphisms, which is a complexity explained through Piaget's idea of *reflective abstraction*. This lesson should change our understanding of structure as a kind of abstract "form." While in this book I have no pretensions to discuss the meaning of structures in other fields such as mathematics, the meaning of a structure in the context of natural intelligence is conceived as an hierarchical architecture of morphisms that are value based, contextual, and unfolding in time. What is a difference that makes a difference for the hen is not necessarily meaningful for the sniffing dog or the curious intellectual. However, underneath the surface, they seem to share a deep similarity of differences.

Summary

- An insightful experiment by Wolfgang Köhler illuminates the relational structure of natural intelligence.
- A difference "makes a differences" only if mapped onto a value/reward category.
- Natural intelligence is grounded in meaning and not in abstract differences per se.
- Natural intelligence is composed of differences, similarities, differences of similarities, and similarities of differences.
- Structures of equivalence and value are evident in various domains, from perception to semiotics.

Chapter 8
From Identity to Equivalence

In Chap. 6, I illustrated the way in which the transformations of an apple as it is eaten may be used to grasp the structure of an apple. This illustrative example invites a deeper discussion about the difference between identity/equality and similarity/equivalence. If you have studied philosophy, you probably know that the notion of identity has been of great concern to philosophers.

Identity may be defined as the relation a thing bears to itself. This definition may be conceived as circular as the thing (which is both the domain and the co-domain and both the source and the target of the identity relation) is what is defined through the identity relation. We may understand the notion of identity in context by recalling that it emerged in our collective mind through classical Greek culture, which was a culture highly immersed in the visual modality and its artifacts, such as paintings and sculptures (Eco, 2000). When you are deeply involved in the production of sophisticated artifacts that mimic the natural and the imaginary worlds, questions of epistemology, representation, and illusion will necessarily pop up, as through your practice you will be involved in and enchanted by the interplay between presentation and representation.

The illusory nature of the visual arts – from painting to cinema – invites reflections upon the gap between representation and reality. This is evident in cases such as the hysteria evoked among the audience of the first film presented by the Lumière brothers, in which a train seemingly running straight at the cinema screen toward the audience was conceived as a real train.

© Springer International Publishing AG 2017
Y. Neuman, *Mathematical Structures of Natural Intelligence*, Mathematics in Mind,
https://doi.org/10.1007/978-3-319-68246-4_8

One possible explanation for the intensive efforts of Greek culture to identify the common denominator underlying the flux of appearances (efforts evident in, for instance, the invention of the Platonic forms) is a desire to resolve the painful uncertainty associated with the epistemic experience and an endeavor to establish a secure anchor for epistemology. However, and speaking *sotto voce*, one must admit that the identity solution, when taken seriously, is no less of an anxiety buster than the Heraclitian *panta rhei* it aimed to resolve. Therefore, sameness should not necessarily be limited to the notion of identity, and, as described by Mazur (2008), even in the context of mathematics, it may be better to substitute the idea of identity/equality with that of *equivalence*.

This point can be illustrated through the apple example presented in Chap. 6, specifically through the images of the fresh apple (to the left of Fig. 6.1) and the almost totally eaten apple (to the right of Fig. 6.1). An intelligent child observing the apple being eaten step by step would have no problem in acknowledging the "sameness" of the apple along its consumption process. However, the presentations of the same apple at the beginning and at the end of the consumption process are *not* identical. They are not even equal, in the sense that a smart child wouldn't exchange the fresh, uneaten apple for the almost totally eaten apple regardless of any philosophical argument concerning identity, equality, or Platonic forms. Plato himself couldn't have persuaded a smart child to exchange the fresh apple the child holds for the eaten apple Plato has just consumed, based on the existence of an underlying abstract "Platonic" form for which the concrete apple is only a limited and poor reflection.

The apple is not identified through the identity function but through some kind of equivalence formed through its changing representations and their value-based mappings. Similarly, the identity of a person cannot be established by seeking an identity function between the "self-object" and itself. Our sense of coherence and relative stability across time and context (similarly to the "identity" of the eaten apple) is constituted through morphisms at various scales of analysis and their value-grounded anchors in memory.

What do we actually mean when we talk about equivalence? In this book, I have replaced objects with categories and relations with morphisms, so when we discuss equivalence, we are discussing the *equivalence of categories*. Let's consider two categories signifying the fresh apple (A) and the eaten apple (E). The two categories are considered equivalent if there is a functor (F) that maps A to E and an inverse functor (G) that maps back from E to A. See the next figure (Fig. 8.1):

Fig. 8.1 Functors between the apples

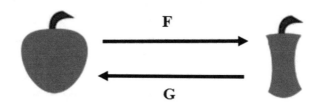

The composition of the functors (i.e., G following F) does not necessarily have to be the identity map but a "softer" version, suggesting that each object of F will be *naturally isomorphic* to its image under composition. In other words, when we map A to E and back again, the functor G doesn't have to return us to exactly the same object from which we departed; we may instead be satisfied by returning to points *isomorphic* to those from which we departed. There is a lot of sense to this, as isomorphic points are indistinguishable, and therefore it seems like a good deal to save effort while returning almost to the same place from which we departed. We can conceptualize this situation as flight paths between two countries. You may pay for a ticket to fly from city x in country C to city y in country D, but on your way back you return to city z in country C. City x and city z may be the same distance from your hometown, and the fact that you return to city z may save you some money. In this case, the fact that you've not returned to your exact point of departure entails no loss of time but is economically justified.

To further explain the idea of equivalence, let me introduce the concept of *natural transformation*. We start our explanation by having two categories (C and D) and two functors originating at C and ending at D. This means that there are two distinct ways in which C is mapped to D. For example, we can think of these mapping functions as two possible translations of the same text to another language. The natural transformation involves the movement from one translation to the other while respecting the internal structure of the original text. To continue with our example, we may translate the text from, let's say, Hebrew to English in two different ways that may reflect different stylistic, lexical, and grammatical choices of the translator. In both cases, however, we must respect the internal structure of the original text; otherwise, our translation wouldn't count as a good translation.

A natural transformation, mapping translation F to translation G, is such that we may move between the first and the second translations while still being loyal to the structure of the original Hebrew text. Metaphorically, we may say that the functors F and G provide different "pictures" of C inside D (Goldblatt, 1979, p. 198) and that the functor between the two functors (i.e. F → G) allows us to see one translation in light of the other. For example, when trying to understand myself during a psychodynamic therapy, I may observe in my self two different "pictures," the one which is my self as a child and the other one is my self as a mature man. Interestingly enough, the different pictures of C inside D may be similar in the sense that the translations forming them may be somehow translated from one to the other. Following my previous example, during a psychodynamic process of psychotherapy, I may try to resolve the discrepancy between my self-image as a child and my self-image as an adult by observing one in the light of the other. In Fig. 8.2, you can see the two categories C and D, the two functors F and G, and the functor mapping functor F to G:

Fig. 8.2 A functor
between functors F and G

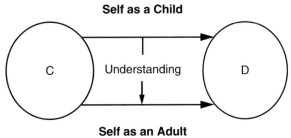

In this figure, the functor mapping functor F (i.e., self as a child) to functor G (i.e., self as an adult) is actually a process of understanding. Let's return to the more general and abstract perspective. To reach the translation from one functor to the other, for each C-object (*a*), we assign an arrow in the category D from the F image of *a* to the G image of *a*; denoting this arrow as τ, we get:

$$\tau_a : F(a) \to G(a)$$

Let's explain this point by using the above psychological example. Recalling myself as a child, I hold an object *a* which is the observing "I." I also have the representation of my father as object *b*. The map between I and father is a relation involving anger. We may translate the word *I* in category C to the same I in category D and find that the image of my father is translated to the representation of my boss with whom I have the same relation of anger. Therefore, in the D category, we should have an arrow:

$$\tau_a : I \to \text{Boss}$$

However, to ensure that this process is structure preserving, we also ask that each C arrow f: a → b is embedded in Fig. 8.3, which commutes:

Fig. 8.3 Natural transformations

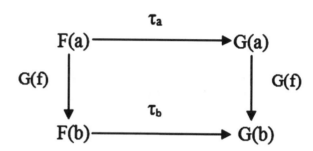

The natural transformation from F to G is therefore a procedure τ that assigns to each object in the C category a D arrow in a way that commutes, as can be observed in Fig. 8.3. The natural transformation τ involves arrows τ_a, which are called the *components* of τ. If each component of τ is an iso arrow in D, it may be interpreted that the F-picture and the G-picture of C look the same in D, in which case τ is described as *natural isomorphism*. Goldblatt (1979, p. 198) suggests that "a reasonably intuitive idea of 'transformation' from F to G comes if we image ourselves trying to superimpose or 'slide' the F-picture onto the G-picture." In our psychological example, trying to resolve the different pictures of myself as a child and as an adult can therefore be described as a process of natural transformation. Through this process, I may understand that my difficult relations with my father are extended to different types of authority figures, like my boss, and that this unresolved conflict with an internalized authority figure (i.e., my father) hinders my personal development.

Now we may return to our apple example as this is a simpler and concrete example. Assuming that the three successive appearances of the apple are appearances of the *same* object, we may consider the two-apple and three-apple versions of the pictures of the apple as two different pictures of Apple-1 inside a new category composed of Apple-2 and Apple-3. See Fig. 8.4, in which Apple-1 is on the left and Apple-2 and Apple-3:

Fig. 8.4 Two mappings
from the first appearance
of the apple

The two apples in the right-hand category may actually be conceived as two pictures formed by two different functors from Apple-1 (to the left) to the apple category to the right. The picture of the apple that appears at the top right may be considered the picture formed by functor F, and the picture of the apple that appears at the bottom may be considered the picture formed by functor G. In fact, the difference between Apple-2 and Apple-3 in the right-hand category is just one bite taken from the apple.

We therefore may consider the transformation from the first appearance of the apple to the following successive appearances as expressing the idea of natural transformation, as there is a way of moving from the first transformation to the second transformation while preserving the original "meaning" of the apple.

Here the notion of *category equivalence* becomes involved. If the functor F from C to D has an inverse functor G, then for an *a* object of C, we have:

$$a = G\big(F(a)\big)$$

and for *a b* object of D we have:

$$b = F\big(G(b)\big)$$

However, equality is sometimes too difficult to attain, and we may settle for isomorphism ($\cong$), which means that:

$$a \cong G\big(F(a)\big)$$
$$b \cong F\big(G(b)\big)$$

This will happen when:

$$a \rightarrow G\big(F(a)\big)$$

and

$$b \rightarrow F\big(G(b)\big)$$

are natural!

As explained by Goldblatt (1979), a functor F denotes category equivalence if there is a functor G such that there are natural isomorphisms τ: $1_C \cong G \circ F$ and σ: $1_D \cong F \circ G$ from the identity functor on C to G ∘ F and from the identity functor on D to F ∘ G. Therefore, the categories C and D are equivalent when there is an equivalence F: C → D.

We may explain this notion of equivalence through the important concepts of *adjointness* and the *adjoint functor*. Let's return to Köhler's hen example, introduced in Chap. 7. We have two categories: the order relation between light gray and gray and the order relation between light gray and white. Let's signify these categories as L and D, respectively. We have two categories and functors (F and G) between them, enabling an interchange of objects and arrows as in Fig. 8.5:

Fig. 8.5 Adjointness

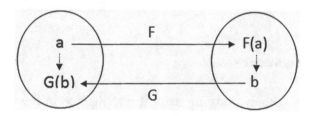

As explained by Goldblatt (1979), adjointness occurs when there is an exact correspondence of arrows between these objects in the directions indicated by the broken arrows in Fig. 8.5, so that any passage from *a* to G(b) in C is matched uniquely by a passage from F(a) to *b* in D, as follows:

$$a \rightarrow G(b)$$
$$F(a) \rightarrow b$$

the identity mappings from each object in C and D are formed as two natural transformations, as indicated above:

$$1_C \to GF$$

$$FG \to 1_D$$

This mapping is described as "natural" as it supposed to preserve the categorical structure, as a and b vary. See Fig. 8.6:

Fig. 8.6 Adjointness in Köhler's hen experiment

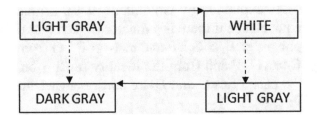

Here, a is the top left light gray, $F(a)$ is white, b is the bottom right light gray, and $G(b)$ is dark gray. What is fascinating about the above (left) "adjoint" is that the map between the two categories doesn't end with identity! It is not an isomorphism, and we don't land at the exact point from which we departed. This is what makes the idea of adjointness and the adjoint functor such central concepts for understanding the behavior of the hen in particular and the logic of natural intelligence in general.

Summary

- When studying natural intelligence in the wild, we should substitute identity and isomorphism with equivalence and similarity (respectively).
- Equivalence of categories may provide us with the appropriate alternative to the limiting notions of identity and isomorphism.
- The concepts of natural transformation and adjointness may further enrich this alternative.
- Structures are formed by identifying equivalence between categories at various scales of analysis.
- This conceptualization is applicable to studying natural intelligence in various contexts, from the behavior of Köhler's hen to the human use of metaphors.

Chapter 9
On Negation

In the previous chapter, we deepened our understanding of category equivalence, which may enrich structures. As the thesis put forward in this book is not presented in a linear order, in this chapter I allow myself to take a step backward to the basics by digging deeper into the meaning of a *difference*. This temporary drift from the main narrative is important for understanding later chapters, specifically for understanding how the concept of zero is represented in our mind. Let us start by reviewing some of the ideas discussed in the previous chapters and by returning once again to Köhler's hen.

In the experiment, the hen responded to a difference that had turned into a difference that makes a difference when mapped onto the value/reward category. The difference that triggered the hen's response was the relation of order, which is a one-dimensional relation albeit not the most simple and primitive relation. The most primitive relation is that of a binary difference between two categories only. It is the difference between nothing and something, between 0 and 1, and it is a difference that has triggered the imaginations of many thinkers, such as George Spencer-Brown (1994). The difference between nothing and something can be represented as follows:

$$\{\ \} \rightarrow 1$$

In categorical terms, it is an arrow from the initial object (signified as {} or 0) to something that is necessarily the terminal object in this two-object category. Something emerges out of nothing. What is the meaning of the arrow originating at the initial object (e.g., the empty set)?

© Springer International Publishing AG 2017
Y. Neuman, *Mathematical Structures of Natural Intelligence*, Mathematics in Mind,
https://doi.org/10.1007/978-3-319-68246-4_9

One possible answer to this philosophical question comes from an unexpected source, which is a short and insightful paper written by Sigmund Freud and titled "Negation" (Freud, 1925). *Negation* is the most primitive relation. In a two-object category, it simply means that 0 is defined as not being 1 and 1 is defined simply as not being 0. In this world of abstract and primitive differences, objects have no meaning whatsoever except for being differentiated. They are just theoretical geometrical points, and indeed in the language of category theory, a point is defined as an arrow from the terminal object to an object of the category. In this sense, the meaning of the two most basic "elements" (0 and 1) is defined by an arrow from 1 to 0, which is actually a negation of *what is* that defines *what isn't* and which thus recursively defines its source.

Freud (1925) carefully observed the embodied nature of negation in the sense that the basic differentiation is formed when as babies we make a decision about whether to consume an item of food or throw it away; we thereby decide "whether it is inside me or outside me." Freud hypothesized that the original pleasure-ego, which we may equate with the reward function, "tries to introject into itself everything that is good and to reject from itself everything that is bad." Moreover, Freud argued that, at the most primitive stages of thought, "what is bad" (i.e., "alien to the ego") and what is "external" are "identical." These hypotheses imply that the abstract philosophical notions of "good" and "bad" are basically embodied – that is, grounded in our sensory–motor activity and in a reward function.

For Freud, negation is not primarily a logical operator but an *epistemological* operator constituting a boundary between the inside and the outside according to the "pleasure principle." It is primarily a *presymbolic* act. While the logical NOT is a reversible operator, the epistemological operator of negation is an irreversible operator forming a boundary, as it is value laden.

When a baby throws away bitter food he has tasted, this act of negation has no symbolic meaning although it is loaded with value. Only when the act of negation gains a symbolic status does it free itself from its concrete embodiment; at this point, thought is endowed with a first degree of independence. As we learn to shake our head to communicate refusal, as we learn to say "No!" and thereby negate, a boundary is formed and our thought moves toward a higher level of abstraction, rationalization, and imagination. For instance, when in some of the monotheistic religions God is described as good, or the ultimate expression of goodness, the first-order relation of good is associated in our mind with pleasure and the primitive wish to let it in to be united with us. Freud explains this by showing that *affirmation* is a "substitute for union." Indeed, there are many expressions of the endeavor to unite with the things we conceive as good, such as our lovers or God, and these conceptualizations are often described in metaphorical terms using language usually reserved for tasty food. These expressions are interpretable from a structuralist's perspective.

Previously, I have mentioned the fact that, in category theory, a point is defined by an arrow from the terminal object to an object of the category. In the context of the most basic difference category, in which a binary category is formed through negation, we may understand why "point" is considered as the most basic primitive. However, this point is an enriched category in which there is a map from the basic difference to the value category. That which is negated is "outside" and "bad." This definition of the point invites speculation about the "geometry" of the mind. If a point is the most basic difference category formed through negation, how can we analogically define a *line* and a *distance*?

A line can be simply defined as a morphism between differences/points that establishes some kind of similarity up to the level of isomorphism, but defining a distance may be a much more tricky issue. Let's assume that we would like to compare the distances between a and d and between b and d. Given $f: a \rightarrow d$ and $g: b \rightarrow d$, we say that $f \subseteq g$ if and only if there is an arrow $h: a \rightarrow b$ such that the following diagram commutes (Fig. 9.1):

Fig. 9.1 The sub-object diagram

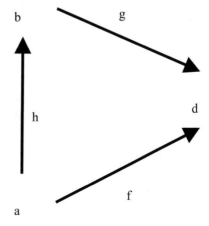

This means that $f = g \cdot h$. If f is equal to g following h, then it means that a is a *sub-object* of b (and d) – for instance, in the case of a taxonomy where a stands for Cat, b stands for Mammal, d stands for Animal, and the morphism stands for "is_a." Now we can say that the distances between a and d and between b and d are the same if there is a map h from a to b but also a map i from b to a. In this case, we may consider f and g as isomorphic sub-objects. Otherwise, the distance between a and d would be longer than the distance between b and d as it factors through b and not vice versa. That is, from the basic negation difference, we have been able to define the analogies of a line, and even a nonmetric distance.

In fact, we may further this idea through the notion of the *simplicial complex*, which is a set composed of points, line segments, triangles, and their *n*-dimensional counterparts. A point is 0-simplex, a line segment is 1-simplex, a triangle is 2-simplex, and so on. When we form a two-unit structure in which the morphism is an isomorphism, we actually form a 1-simplex, which is a groupoid. When we add to this groupoid a third component, we have a triangle, which is a 2-simplex. This idea can be illustrated through semiotics, where each sign (e.g., word) is a basic difference forming a point or a 0-simplex. Now, in the case where two objects factor through each other with regard to a third object, we may consider them equivalent and as forming a resemblance and thus a line between them, which is a 1-simplex, indicating similarity. This may be illustrated with regard to the equivalence relation formed between cherry and strawberry with regard to a third object fruit.

The fact that a "line" may be formed between two objects with regard to a third might be considered meaningless, but, for semioticians familiar with the work of Peirce, the idea that meaning is always formed within a triadic relation, always with regard to a third, may be another expression of the deep logic underlying natural intelligence. This kind of "geometry" grounded in a structuralist approach to natural intelligence may be of relevance for various models of natural computations, but at this point, it remains at the level of an intellectual exercise only.

Now let's return to the basic binary notion of difference. I have described the basic binary notion of difference as composed of two well-demarcated values: 0 and 1. This basic description says nothing about the nature of categories in the real world. Freud (1925) himself argued that thought is basically an "experimental action"; therefore, imagining the basic categories as 0 and 1 should not be judged as an ontological statement about the way things really are but as an expression of *judgment*, which is "the intellectual action which decides the choice of motor action." In other words, the structure of the most basic difference category and its "sharp" binary values is explained through the operation of natural intelligence, which has to judge (i.e., to choose) between two competing courses of action. When an infant playing in the yard tastes what seems to be an appealing bug and finds that it tastes awfully bitter, she will almost instinctively spit it out of her mouth. This action doesn't express an oversimplified epistemological statement that ignores the complex and gradual nature of bitter taste. It is a simple expression of a judgment motivated by the need to avoid the consumption of poisonous materials.

At the most basic level, judgment is grounded in binary decisions, such as to eat something or spit it out. This basic logic of natural intelligence may explain why information theory, which is built on binary choices, is so appealing to us in studying living systems, despite the fact it was originally developed for studying artificial systems. Having said that, we must of course admit that, in the real world, information and differences are stretched along a physical continuum, and therefore natural intelligence has to deal with judgment along a continuum.

For example, *chemotaxis* is a process that involves the movement of organisms, such as bacteria, in response to chemical stimulus. Bacteria sense their environment and move toward regions rich with food while moving away from poison. Bacteria face the challenge of mapping a chemical gradient into a two-value difference category, a process that has been documented and formalized in relation to various natural sensors with relevant mathematical solutions (e.g., the sigmoid function). However, the bottom line is that, for bacteria, the world may be composed of a basic difference category of food vs. poison, which in turn is mapped onto a reward category that forms a difference that makes a difference, from which derives the observed action.

The logic of bacteria thus may seem very simple: randomly swim in your environment and sense what surrounds you. When you sense food, move toward it. When you sense poison, escape from it.

Otherwise, keep moving. In practice, this behavior is much more complex as it may involve communication between the bacteria and their emergent behavior. However, at its heart, the logic seems to be very simple: a basic difference category formed through membrane receptors, a basic reward category associated with the difference category, and the resulting behavior, which is movement toward or away from a chemical stimulus.

At this point, we may understand why the behavior of the hen seems to be much more intelligent than the behavior of a bacterium. The bacterium's mind may be conceptualized through a single and quite simple difference category. In contrast, the hen has shown the ability to think in terms of *similarity of differences*. It has formed the analogy light_gray:dark::white:light_gray. The similarities of differences formed by the hen is an isomorphism between the two difference categories. This similarity may be the basis of *clustering*, as "objects" holding the same positions in the difference categories may be equated. For example, if Danny plays the guitar and Diana plays the piano, then we may group guitar and piano into a single category of musical instruments. However, the similarity of differences may next produce a difference of similarities, as playing the piano may be quite different from playing football.

The hen may develop the category of predators, which would include all possible objects threatening to annihilate the hen, from foxes to cars. However, and despite its impressive cognitive capacity, the hen may be comparatively limited in forming similarities that don't adhere to the isomorphism type of similarity. In contrast, human intelligence has the potential to operate on a wider spectrum of differences, to form more complex categories (which form high-dimensional structures), to identify and create similarities based on equivalence, and to construct similarities of differences and differences of similarities at increasingly higher levels of abstraction.

For example, we may think about connotations and metaphors as emerging from the equivalence of categories combined with a forgetting function. The general structuralist recipe for such an activity may be as follows. When your mind builds a basic category such as lawyer, you may identify similarity of differences between lawyer and shark by inventing the metaphor "my lawyer is a shark." I will further elaborate on these ideas in the following chapters.

In sum, I have followed Piaget's structuralism but sought to elucidate the abstract characteristics of a structure using the language of category theory. I have also identified the *groupoid* as a building block of structure and proposed that the rigid isomorphism criterion may be "softened" in favor of an equivalence-based kind of similarity. I've also proposed that the groupoid may have a dynamic nature formed through the synchronization or association of objects. Moreover, I've argued that, as the groupoid forms an equivalence group, it may be used to cluster lower-level elements and *abstract the produced clusters* for processing at a higher level of analysis. The anecdote of the hen has been used to explain how natural intelligence is based on relations/differences and the formation of similarities of differences and differences of similarities, while forming meaningful units that (following Bateson) I've described as a difference that makes a difference. I've also emphasized the contextual and value-based aspects of natural intelligence and that the structures underlying natural intelligence emerge from these contextual and value-based episodes. The minimalist structuralist program that I've presented so far may be further elaborated and refined through case studies.

Summary

- The most basic difference category consists of the relation of negation.
- It is a difference in the binary category that includes two objects: 0 and 1.
- In contrast with the reversible logical operator *not*, negation is an epistemic and irreversible operator that forms a boundary.
- Through this difference category, a point can be defined as a map from the terminal object.
- This idea explains the "geometry" of mind and the way clusters of objects are formed and produce abstract categories.

Chapter 10
Modeling: The Structuralist Way

Natural intelligence involves the attempt to model "reality," and I have approached this modeling process by using the general aspects of structure. Setting aside philosophical sophistication, the modeling process simply means that organisms of totally different natures are involved in responding to a *mediated* reality – that is, to *re*presentations of information they gather from outside the system as well as from inside the system. This idea has been philosophically supported by Peirce and seems too trivial to be either defended or justified today. The idea that organisms "model" reality is simply deduced from the fact that they cannot operate directly on their environment and are *by default* "doomed" to acknowledge that they have only maps and that the map is not the territory it represents.

In practice, the modeling of reality is a distributed process in which a vast network of interconnected agents orchestrate their activity in order to integrate a variety of signals into a meaningful representation on which actions can be based. Therefore, when we actually model the behavior of a complex modeling system, such as the immune system or the human brain, we usually do so by focusing on a limited spectrum of this modeling activity and by using a simplified model of the modeling process. In other words, academics striving to understand natural intelligence are involved in second-order modeling that must be communicated between human beings and therefore has to be much simpler than the process it models.

Think, for example, about the complex coordination involved in a starling flock (e.g., Attanasi et al., 2014). Each bird must coordinate

© Springer International Publishing AG 2017
Y. Neuman, *Mathematical Structures of Natural Intelligence*, Mathematics in Mind,
https://doi.org/10.1007/978-3-319-68246-4_10

its flight with the flight of others in order to avoid midair clashes. This complex process (materialized in each bird's brain) is a modeling process, as each bird must make decisions based on the way information about the loci of other birds is represented in its brain. However, when scientists are modeling this process, they significantly simplify it in order to reach some general conclusion. Natural intelligence is mostly unconscious and doesn't bear the burden of communication. The starlings have not developed theories about flying in a flock, they don't bear the burden of communicating a theory of flight patterns with (say) pigeons, and they don't have to justify their flying empirically in academic conferences with other birds. Their brains focus on the activity of flying rather than on its reflections and they attune their parameters on a "sub-symbolic" level to gain optimal performance, as may be the case with other learning processes, such as those described in the context of machine learning (e.g., Battiti, Brunato, & Mascia, 2008).

It is therefore argued (Battiti et al., 2008) that one of the challenges facing natural intelligence in the processes of modeling, learning, and problem solving is to find the appropriate balance between exploration and exploitation, elsewhere defined as intensification and exploitation or diversification and exploration.

This balance may be illustrated in several contexts, such as cockroach escape behavior, human personality, and human immune recognition. We will explore each of these in turn.

First, let's assume that, as non-blind watchmakers, we would like to optimally program the escape behavior of a cockroach. Given a threat (e.g., an approaching human being), the cockroach has to distance itself from the threat by finding the shortest path from its current location to a dark place far from the light (where it is an easy target). The intuitive heuristics of the cockroach may be to move in a straight line to a locus where the boundary between light and dark is maximal, signaling a potential hiding place. However, moving in a straight line in order to minimize the traveling distance or time between two points is a conservative albeit a risky move, as movement in a straight line makes it easy for the human to predict the bug's next move and exterminate it. Therefore, the cockroach may adopt a strategy that combines a target-oriented movement with random shifts that add the vital ingredient of unpredictability to its motion. While escaping from the threat, the bug is searching the space for solutions (e.g., escape paths) that might solve its problem (e.g., escaping from a potential approaching threat). Adding a random component to the movement is exploratory in the sense that it makes many more escape paths available, lowering the predictability of the bug's movement, but also entails a higher risk, resulting from the extended exposure time to the threat and from the fact that random movement may even bring the cockroach closer to the threat.

A similar dynamic may be illustrated in the context of the human personality. Let's assume that you are a member of a hunter-gatherer people and have an introverted personality and a low openness to experience. By chance, you stumble upon a group of mango trees. Being rewarded by this rich and exclusive source of food, you form a settlement at the group of trees, deciding that you've solved the problem of finding food. Focusing on this simple solution is rewarding in the sense that food is secured with almost no risk. However, in nature, being too secure is a risky strategy. Settling by the mango plants doesn't give you access to a wide spectrum of nutrition. For instance, the mango can't provide you with proteins, which can be gained through hunting. Hunting is of course a more exploratory solution involving a high level of risk. Those who hunt will gain access to proteins but with more effort and with the risk of wasting energy and confronting animals.

In a free and competitive "market" such as nature, being a conservative investor means being at almost as high a risk as being an adventurer investor. A balance therefore has to be found, and different personality types express these chosen preferences in different ways, combining being open to experience (at the price of risk) and being limited to minimal solutions (at the price of missing opportunities).

The same is true of our immune system, which is responsible for recognizing pathogens. A conservative approach, epitomized by the innate and more basic part of the immune system, is to develop general templates for recognizing pathogens. When something is identified as having general structural markers of "bad guys" or as lacking a marker of the self, the system attacks it. However, being conservative means being at risk as the evolution rate of pathogens is such that the old templates may not keep up and thus many "opportunities" for identifying pathogens might be missed. In contrast, in the adaptive and more sophisticated part of the immune system, many immune "sensors" are produced, parameters are attuned, and decisions are reached through sensitive contextual analysis (Cohen, 2000).

The cases that I've presented above represent the *dominant* approach to the modeling of intelligent systems. An intelligent system builds a model of the environment. The model may be composed of a fixed skeleton of variables and their parameters in which the parameters are attuned such as to *optimize* the model's performance against real-world criteria, such as finding a balance between exploration and exploitation in the search for food. In sum, the dominant modeling approach assumes that the main challenge of learning is *optimization*.

This dominant approach to modeling may be enriched by the old Indian story about a group of blind men and an elephant. Several blind men touch an elephant to try to learn what an elephant is. As each of them touches a different part of the elephant, they finally reach the conclusion that they are in total disagreement and become aware of their ignorance. Becoming aware of our ignorance was an important phase in the development of science as it led to the formalization of the concept of ignorance, as epitomized in the emergence of probability theory. To explain this point, we must better understand the concept of *chance* and the way it has been understood and formalized (Lüthy & Palmerino, 2016).

The concept of chance is used to refer to surprising and unexpected events, or events that seem inexplicable. In the first case, chance is used to express a violation of our expectations, such as when I meet a close friend when traveling abroad while holding the belief that he is back home. The second sense is a way of expressing our ignorance of the causes that led to an event (e.g., "Why did it happen to me?").

Darwin used the concept of chance in the second sense, as he could not explain the cause of the variety he observed among species. Even today, although we better understand genetics (which is responsible for phenotypes) and can talk about "random mutations," the way this "randomness" is generated remains far from clear. The idea of chance as an expression of ignorance suggests that chance is therefore best understood *ex negativo*, or from our ignorance (Lüthy & Palmerino, 2016), and that there are various ways of conceptualizing modeling *ex negativo*, one of which will be presented in this chapter.

Chance also has the sense of being structureless. The logician and philosopher Per Martin-Löf's notion of randomness suggests that a sequence is random if it cannot be compressed, meaning that it has no underlying structure that enables its representation in a more parsimonious and economical way. This idea of chance or randomness echoes the ideas I have already presented in this book, as the ability to map/ translate a sequence onto an ordered representation is a specific case of structure-preserving mapping. Indeed, something is surprising when it violates our expectations that it will appear to have a known structure.

In a horror movie I watched many years ago, we see a child sitting in front of a TV. The audience's perspective is located behind the TV in such a way that only the upper half of the child's face can be observed. When the camera moves downward, we realize in horror that the child has no mouth. The horror is the feeling accompanying our negatively loaded surprise resulting from the violation of the normal face structure. The "uncertainty" involved in this case is not the same as the uncertainty we conceptualize in probability theory, as it is not expressed in the known and equal frequency of events or in our subjective expectations, although one may easily translate the above situation to Bayesian expectation about the existence of a mouth in a child's face. However, the statistical interpretation of chance, whose evolution has largely been inspired by games of chance, is suitable for handling systems of low dimensionality in which the elements' respective behaviors are independent of each other, such as in a container of gas molecules.

Probability theory is also a theory about *sets of objects* and not about individual objects (Terwijn, 2016). However, when modeling high-dimensional and multilevel systems in which relations and structures are prevalent, interpretations of probability relating to uncertainty and chance cannot be easily scaled up as they are grounded in and limited to sets, whereas categories impose some structure on sets. This should not, though, be interpreted as an attempt to dismiss the enormous achievements of probability theory or information theory; the aim is only to point to their limitations and open the way for new ideas.

In addition to the above criticism, we should recall that organisms encounter individual objects/structures and not mathematical sets. The "chance" expressed in the horror movie as described above results *ex negativo* from the violation of a structure-preserving map. Our expectations always have a domain and a co-domain, a source and a target. The surprise is therefore a surprise grounded in structural expectations associated with some concrete value judgment.

As we can see, the naive concept of chance is embedded in complex issues, such as the nature of expectations and the meaning of causality. In various fields, human beings have realized that the certainty of simple mathematical models is incomplete and that uncertainty/ignorance is built into many real-world processes. Instead of trying to "solve" this uncertainty, the brilliant epistemological move of probability theory was to *define and formalize* uncertainty in a way that can make it useful. In other words, probability theory didn't solve the problem of uncertainty by turning it into certainty but "domesticated" it by giving a form to chance, randomness, and uncertainty.

The concept of uncertainty, though, has various senses, and the story about the blind men and the elephant seems to introduce a notion of uncertainty quite different from that proposed by probability theory. In the context of the blind men, the uncertainty concerns the *incongruent* morphisms of the different blind men who produce different pictures within the same "elephant" domain. In other words, the ignorance expressed by the blind men results from the fact that, given their expectation that they are examining the *same object*, their different pictures or models don't seem to be translatable to each other. What can we learn about this kind of uncertainty and what can we learn from these different pictures?

The idea of natural transformations may point to an interesting direction and in fact a novel approach to modeling uncertainty in natural intelligence. For any given phenomenon we are trying to model, we don't necessarily have to find the best parameters or the most parsimonious description (e.g., minimum description length), at least for the first phase, but we may be interested in measuring *the extent to which the different translations are translatable to each other*! This idea doesn't dismiss the uncertainty expressed in statistical learning but rather add another layer to the complexity of modeling and understanding the world.

What can we learn from the incongruence of the different pictures provided by the blind men? Regardless of "what there is" – the elephant – we may learn that incongruent pictures express different aspects of the observed reality. These incongruent pieces may be different parts of a single whole, as in the case of the Indian story of the elephant, or the different expressions of the *same object* over time, as illustrated in the example of the apple (see Chaps. 6 and 8).

The incongruence may also teach us that, regardless of our beliefs, we are not dealing with the same object but with different objects. In sum, the use of natural transformation may be a powerful tool to examine whether:

1. We are experiencing the same object.
2. We are experiencing different parts of the same object.
3. We are experiencing different appearances of the same object over time.

This is a very powerful idea that may support the generation and testing of hypotheses without probability theory! The idea that the laws of probability theory are somehow encoded in the mind may not be the first hypothesis we want to generate. It is more reasonable to hypothesize that minds, as relational engines, attempt to match the various maps they contain. Think about the hen, for instance. If the mind is a kind of probabilistic engine, then the hen's mind is different from the human mind in its level of probabilistic sophistication. It is as if the hen's mind is equipped with the (unconscious) statistical knowledge of a high school student and the human mind with that of an advanced student. However, this is an assumption that we may find difficult to accept, for obvious reasons. It is much more reasonable to accept the neo-structuralist perspective presented in this book and its derived notion of uncertainty.

Up to now, science has considered statistics and information theory as the main perspectives for modeling the way in which organisms model the world. For example, let's assume that you would like to teach a computer to recognize pictures of squirrels. First, you collect a set of squirrel pictures. This set must be representative in the sense that, if you can't provide the computer with the whole population of squirrel pictures, then you must provide it with a representative sample of pictures that covers all types of squirrels, with their different colors, types of fur, ages, genders, and postures. You also provide the computer with a set of other mammals that might be mistakenly recognized as squirrels (e.g., mice). These representative sets are then fed into the computer as digital images composed of pixels in such a way that each digital image is represented as a vector of pixels constituting the "features" that the computer should use to classify the images as squirrels vs. non-squirrels.

The problem is now the classification problem we have discussed before, and the uncertainty is expressed as the probability of the image being classified as squirrels vs. non-squirrels given a specific set of features and their values. As the features' combinatorial space might be astronomically large, finding a classification function that provides the optimal solution to the classification problem turns out to be a problem of optimization and the automatic fine-tuning of the parameters in the classification function/model.

In contrast, the modeling process that I've proposed is totally different as we are not interested in providing an optimal solution by fitting the model to the data but in (1) examining the extent to which different models of the world are translatable to each other through the mechanism of natural transformation and (2) generating and testing relevant hypotheses derived from this "translatability." As such, we don't compare the models to determine the relative qualities of different statistical models, as in the Akaike information criterion, and we don't seek the most parsimonious way of describing the data, as proposed by the minimum description length principle. We examine how translatable are our different translations and the lessons we may learn from their mutual translatability. We may describe this idea as the *natural transformation modeling* (NTM) principle.

According to this new principle, we may read the Indian story about the blind men and the elephant in a new light. This story should not necessarily be concluded with a sense of failure but with the potential for better understanding resulting from our ability to produce limited pictures of the world *ex negativo. Ignorance and failure may be illuminating if appropriately leveraged.*

The NTM principle suggests that the uncertainty characterizing natural intelligence results from our basic nature as "semiotic machines" who work in a mediated reality and as a result produce varying and potentially incongruent pictures of the world. This is a crucial point. Given that natural intelligence is mediated, different people *necessarily* form different mappings or functors between reality and the mind. As we are never sure about the "true" nature of this reality, which is always mediated, and in contrast with the procedure used in machine learning, we cannot have a solid benchmark against which our models can be simply tested. As a pragmatic and natural solution grounded in ignorance and hypotheses, the NTM principle proposes to examine the relations between the different maps we form of the world regardless of our understanding of "true" reality as it exists behind the curtain of mediations. It also proposes that we may learn a lot, *ex negativo* and through the formation of hypotheses, from the incongruence of maps rather than from their coherence.

We may of course have different worldviews that are more or less internally consistent but that have nothing to do with reality itself. That is, our models may be reliable and in agreement with each other but, as in a consensus of fools, have no validity. However, and to recap, our maps make sense as long as they may be mapped to a value/meaning category. In contrast, in some of the classical sophists' ideas, our *ex negativo* form of understanding must be somehow connected to a value/meaning category, which ensures that our understanding doesn't remain in the abstract and indifferent realm of pure differences. That is, the difference categories formed by the *ex negativo* principle make sense only when connected to the *reality principle*.

Moreover, using the NTM principle not only allows us to test the translatability of our different maps but also, when combined with the idea of adjointness, allows us to return to the same mediated reality we are trying to model! Assuming that the world is not simply presented in our mind but represented, then the idea of adjointness allows us not only to examine the different maps of reality we form in our minds (whether our private mind or the collective mind of scientists, for instance) but also to approximate or impose equivalence structures on the hypothetical reality we continuously model.

As that which is behind the curtain of mediation is unreachable, we cannot build isomorphic maps of the real but only a variety of maps that may have some similarity *up to the level of isomorphism. The approximate nature of this intelligent activity explains why creativity and the formation of hypotheses, rather than logical deduction or empirical induction, are the hallmarks of natural intelligence.*

Our ability to identify similarities is partial due to the various maps we produce, and, as our similarity formation is based on equivalence rather than on a strong correspondence between mind and reality, hypotheses and creativity are almost inevitable. This is evident in the formation of novel metaphors (e.g., "life is like brandy") where the metaphor, rather than exposing an existing structural similarity between domains (i.e., life and brandy), *creates* a similarity that didn't exist before. The power of the new principle proposed in this chapter will be illustrated in the next chapter.

Summary

- Natural intelligence "models" the world, as reality is always mediated.
- The uncertainty involved in this modeling process is usually discussed through probability theory.
- Uncertainty may be conceptualized through the neo-structuralist approach as an incongruence of maps.
- The natural transformation modeling (NTM) principle describes this uncertainty.
- The NTM principle may explain how we gain knowledge *ex negativo* (from our ignorance).

Chapter 11
On Structures and Wholes

Knowing the microscopic parts, I cannot infer the macroscopic whole unless I am already familiar with it. (Nicod, 1930/1950, p. 60)

The configuration number is the number of a priori equally probable states that are compatible with the macroscopic description of the state i.e. it corresponds to the amount of (microscopic) information that is missing in the (macroscopic) description. (Von Neumann, 1956, p. 24)

That the same way that the whole is, of course, understood in reference to the individual, so too, the individual can only be understood in reference to the whole. (Schleiermacher, 1999, pp. 329ff.)

Piaget described structures as involving some mysterious property he called "wholeness." This property is evident in various systems, such as natural language, where a sign doesn't have meaning as an isolated sign unless it is part of a wider structural whole through which it gains its meaning. This structural whole is constituted through its signs but also recursively loads them with meaning. This dynamic has been named in hermeneutics, the field that deals with interpretation, as the *hermeneutic circle*, and it is evident in the dynamic nature of language (Danesi, 2003), a point that will be further elaborated.

How do we associate differences or "points" to form wholes? The Gestalt *law of proximity* suggests that objects are grouped together if they are *close* to each other. The meaning of "close" should be clarified, of course, as it involves the notion of *distance* and sometimes of a *metric* that allows us to quantitatively determine how far two objects are from each other. I previously introduced such a nonmetric and categorical interpretation of a distance that makes use of the notion of the sub-object. The Gestalt principles also suggest the *law of similarity*, according to which similar objects are grouped together. Again, the

© Springer International Publishing AG 2017
Y. Neuman, *Mathematical Structures of Natural Intelligence*, Mathematics in Mind,
https://doi.org/10.1007/978-3-319-68246-4_11

meaning of similarity is not self-evident, but we have already adopted the idea of similarity as equivalence of categories. At this point, we may understand that objects and categories may be grouped by similarity, as conceptualized along the lines of the previous chapters, and by proximity, as conceptualized through the nonmetric distance formed through the notion of the sub-object.

Let's try to understand how "points" may be grouped together. For illustration, we will follow a simple example in which we try to elucidate the semantic field of Cat. Searching a lexical database for the words collocated with Cat we may find:

$$Cat = \{dog, mouse, pet, chase, lick, milk\}$$

These words are just *data points* we may identify as associated with our target sign Cat. They are grouped together only because they exist in some proximity to the sign Cat, meaning that they are just words that can be found to appear in the lexical context of Cat beyond a certain statistical threshold of surprise.

At the most basic level, our signs as data points exist "in and for themselves" as entities differentiated from each other and no more. The fact that we have decided to group them as objects in the set we titled Cat results from the fact that they happen to co-occur with Cat in the "same place," by association only. Recursively, we may describe each of these data points by using the set of words with which it is associated. For example, the data point Dog may be described as:

$$Dog = \{cat, owner, pet, walk, eat, bark\}$$

and the data point Mouse may be described as:

$$Mouse = \{cat, rat, gene, catch, infect, inject\}$$

At this point, we have a set of data points each of which can be described by its own data points. Next, we would like to connect the dots. I have proposed that a line may be defined, the *categorical way*, by examining every two data points with regard to a third and asking whether they factor through each other. For example, Dog, Mouse, and Pet are three objects that are associated with Cat, but how telling is Dog about Mouse with respect to Pet? How telling is Mouse about Dog with respect to Pet? By using the Kullback–Leibler divergence measure (see Chap. 3), we may find that Dog and Mouse are more telling about each other than Dog and Pet as they share more words in their semantic fields. We may therefore decide to form a line between Dog and Mouse. We may also find that an additional data point, Hamster, forms a line with both Dog and Mouse, and we may therefore gain a set of three data points clustering objects that are pets. These clustered sets, which are probably *disjoint sets*, are actually groupoids formed within the set we have identified through proximity only. In other words, first we identify objects that are close to each other. The set formed by the law of proximity forms a *boundary*. The boundary defines the set of Pets.

The law of similarity adds some depth and abstraction to the activity of clustering as it requires that some equivalence is formed between objects. At this point, we have a boundary around a "whole" that may be decomposed into sub-objects by forming subclusters. In sum, the process of clustering is actually constituted by first identifying similarities of differences that exist at 0 dimensionality. That is, the mechanism for identifying similarities of differences, at various scales of analysis, is also the one responsible for grouping or clustering. In contrast, the mechanism for identifying differences of similarities is also the one responsible for forming differentiation. Let's elaborate the formation of the whole by analyzing a visual image. Take a look at the following picture of a dragonfly (Fig. 11.1):

Fig. 11.1 A dragonfly

This is actually the shadow of a dragonfly, flattened into an abstract 2D representation. This flattening is highly informative and follows the *law of good Gestalt*, according to which projecting the 3D dragonfly onto its 2D flat representation may teach us a lot while removing unnecessary complexities. (A similar trick was used by Galileo in his attempt to understand motion; Lawvere & Schanuel, 2000, p. 3.) By mapping the dragonfly from space to plane, we form a shadow of the dragonfly. By analyzing this shadow, a child may easily identify the different objects/parts of which the dragonfly is composed.

These parts are the *factors* we may identify in the image by considering it as a visual product of a body and four wings. Through this procedure, the child may identify the symmetry between the two upper wings and the two lower wings. In fact, the child may decompose the dragonfly into different parts by first realizing that the dragonfly can be decomposed into subclusters and then that these parts or subclusters (e.g., the wings) form symmetric shapes. We may also "break" the dragonfly into sub-objects if we follow the sub-object scheme that has been presented before (e.g., Fig. 11.2):

Fig. 11.2 Sub-objects of the dragonfly

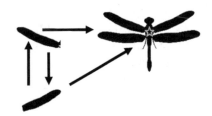

The star at the upper middle part of the dragonfly signifies its center, a point that will be elaborated shortly. The fact that the two wings, as two distinct clusters, can *factor through each other* and be mapped to the whole image turns them into isomorphic sub-objects of the *whole* image. That is, the *recursive–hierarchical* (Harries-Jones, 1995) nature of the whole is revealed through this form of structural mapping. The wings are considered as two sub-objects of the dragonfly. If we imagine slicing up the dragonfly (which is a boundary demarcated shape) into pieces formed by lower-order relations of proximity/similarity, we might find that some of the pieces can be mapped onto the whole dragonfly.

Not every slice of the dragonfly will easily fit back into the picture. For instance, many years ago, when I visited Hong Kong, I noticed that chefs sliced chicken in an "unnatural" way: horizontally and not by decomposing the chicken according to its various natural parts, such as the legs (i.e., drumsticks).

We may identify parts through the use of heuristics, such as identifying regions maximizing boundaries between the parts. The chicken illustrates this point as its legs are connected to its body in places that form natural demarcation points.

Fig. 11.3 The center of
the dragonfly

Let's return to the original image of the dragonfly. The fact that some of these sub-objects can be factored through each other turns them into higher-order structures that can be considered as groupoids. The wings have no meaning without being contextualized within the whole dragonfly, and the dragonfly cannot be comprehended without being understood as composed of its components.

To take this idea further, let's return to the natural transformation modeling (NTM) principle. A child may imagine the dragonfly as a puzzle composed of multiple parts. An attempt to form natural transformations from the functor of each part category to the whole dragonfly category may teach the child that her picture is incongruent. As her first hypothesis, she may wonder whether these are part–whole relations. By sliding some parts above others (e.g., upper wings above lower wings), she may see that they are congruent and therefore form local symmetries. By applying the sub-object morphisms described above, she may complete the puzzle and learn that the components can be comprised into a wholeness. This description isn't far from what children actual experience, as children learn by experiencing with the world.

Importantly, the symmetry formed through the similarity mapping between the two wings, which are local symmetries and hence form a groupoid, also reveals that this structure has a *center* located in between the isomorphic sub-objects. See Fig. 11.3, where the center is marked with a gray star.

This is a key point for understanding the formation of a structure as a whole: groupoids, as "local symmetries," *help us to identify the center of a structure*. In his monumental book *The Nature of Order*, the architect Christopher Alexander (1980/2002) argued that "wholeness" involves center(s), an idea that follows the Gestalt *law of symmetry*.

The center is a hypothetical concept that functions as a point of reference for learning and reasoning. This is a crucial observation. Natural living wholes (or wholeness as described by Alexander) have centers at various scales of analysis. Have a look at the image of a checkerboard in Fig. 11.4:

Fig. 11.4 A checkerboard

Can you identify the centers around the symmetry axes? Can you see the difference between finding the center(s) of this human-made artifacts and the center(s) of the dragonfly? Between a living system and a human-made artifact?

Identifying the center (or the centers) of a structure is highly important for testing the way in which the object is actually composed and for identifying higher-level transformations. Identifying both centers (at various scales of analysis) and centers of centers may be vital for understanding various structures. Think about the anecdote of slicing a chicken. First, we may hypothesize that symmetric objects may be sliced at the point of their intersection with other objects, points of intersection that can be found around the center's axis. The chicken's legs may therefore be sliced off at their point of intersection with the chicken's body.

Along the same lines, a child who has caught a dragonfly may notice that the two wings (i.e., sub-objects) located on either side of the center move in a synchronous way and, if he is cruel and curious, that they can easily be detached from the body at their contact points with the center.

Furthermore, the way in which the different parts of a creature move with regard to each other is another important point in understand the wholeness of structures. When we observe a person walking on two legs, a horse walking on four legs, or a dragonfly flying through the use of its wings, we notice that different forms of transformation exist between the moving parts over time and space. Therefore, the various ways in which the objects of a structure are *dynamically and temporally* related to each other must also be considered. These relations probably include the spatial–temporal transformations that these objects undergo. As such, movement is highly informative about the structure as a whole, which is made up of multiple heterogeneous parts at various scales of analysis.

What differentiates the crystal, the leaf, the human being, and the sophisticated artifacts produced by human beings from each other is not so much the principles of structures but the heterogeneity, level of symmetry, and variability of objects at different scales of analysis as orchestrated in time. At this point, we had better clarify the notion of wholeness.

The famous Gestalt slogan "the whole is different from the sum of its parts" is one of the most insightful psychological observations ever produced. This logic of the whole, or one may say "wholeness," has been interpreted in various contexts, from psychology to hermeneutics and architecture. Here, I would like to introduce a comprehensive and integrative discussion of structure and wholeness along the lines previously presented. This discussion is interdisciplinary in nature as it aims to show how the ideas presented so far resonate with the ideas presented in other contexts regarding isomorphism and how they may enrich our understanding.

My point of departure is the basic argument presented in the first part of this book: when we form a structure or comprehend a structure, whether in architecture or in a sign system, our building blocks are groupoids, here interpreted as *local symmetries*. According to Alexander's 15 properties of wholeness (Alexander, 1980/2002), local symmetries are balanced distributions of forms around some point or line. My suggestion is more general and dynamic as it suggests that groupoids may be formed through some kind of synchronization that creates what Alexander describes as "alternating repetitions." Repetitions, whether within the same object or across time, are key to understanding the general notion of structure. Previously, I've shown how repetitions over time (e.g., the apple example in Chaps. 6 and 8) may lay the ground for the abstraction of apple and how repetitions within an object (e.g., the dragonfly example) lay the ground for understanding the structure of an object.

It is important to emphasize that these repetitions are not necessarily *replications*. When we observe the shape of the apple as it is being consumed along a time line or when we observe a bird flying, we don't assume that the different appearances of the apple or the bird are exactly the same. In fact, this is not even a working hypothesis of natural intelligence, as it could serve to lead natural intelligence astray. Local symmetries, as formalized through the concept of the groupoid, are therefore recognized *ex hypothesis*, as one might say. When we observe an object along a time line (observe its repetitions), we hypothesize or actually construct or impose some order on it, creating a working hypothesis that seems to be necessary to elucidate the structure of the object as it unfolds over time. Moreover, repetitions are not only evident in fields where the visual dimension is dominant. Repetitions clearly appear in music in the form of rhythms and harmonies, and also in language, where the repetition of signs in a text constitutes its *cohesion* (Hoey, 1991) or its perceived coherence, "harmony." Repetitions are formed through gradual transformations of a category. Two categories may be identical, isomorphic, or equivalent. The more we distance ourselves from the original category, the less evident are the repetitions.

As I have shown in the dragonfly example, local symmetries form *centers*, which is another property proposed by Alexander (1980/2002). A structure doesn't have to include a single strong center. In living systems, and in contrast with artificial human-made abstractions or artifacts, we usually observe a variety of centers at different *scales of analysis*. Scales of analysis characterize living systems. This is why "flat" models that don't take into account multilevelness, such as thermodynamic models, are inadequate for understanding natural cultural systems.

A center may be hypothesized to exist between local symmetries and at various scales of analysis. As a search heuristic, we may seek the centers between local symmetries, but we may also think of the center as a *fixed point*. A fixed point of a map is an element of the domain that is mapped to itself. Think about the example of rotating a disk (Lawvere & Schanuel, 2000, p. 121), such as a roulette wheel that turns clockwise (Fig. 11.5):

Fig. 11.5 A rotating disk

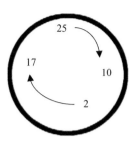

Have a look at the points at the axes of the arrows. They have been arbitrarily signified by "25" and "2." When we turn the disk clockwise, we have actually mapped each starting point from its domain to its position at the co-domain. Signifying the mapping as e, we get $e(25) = 10$ and $e(2) = 17$.

You can easily see that in Fig. 11.5 point 25 has been mapped to point 10 and that point 2 has been mapped to point 17. However, you may notice that when the disk is rotated there is a single point that remains *fixed* and that doesn't change its value. This is, of course, the center of the disk, which remains constant despite the mapping and the transformation. As you can also see, the center may be defined as the fixed point of a giving mapping function. The question is then how to identify these fixed points/centers. One interesting proposal from the field of category theory should be examined, but it requires some definitions.

You will have noticed that in the disk example, the domain and the co-domain are the same object (i.e., the disk). A map in which the domain and the co-domain are the same object is called an *endomap*. An endomap e is called *idempotent* if $e \circ e = e$. Now let's define the concept of *retract*. An object A is a *retract* of B if there are maps

$$A \xrightarrow{s} B \xrightarrow{r} A$$

with $r \circ s = 1_A$.

The endomap e of B constructed by composing r and s in the opposite order $e = r \circ s$ is idempotent: $e \circ e = e$ (for a proof, see Lawvere & Schanuel, 2000, p. 100). This has an interesting property that may be introduced through an example that appears in Lawvere and Schanuel (2000).

In the United States, there are 435 congressional districts. Each district elects a single member as its representative to Congress. In Fig. 11.6, A is the congressional districts and B is the set of people in the United States:

Fig. 11.6 A diagram of
US elections

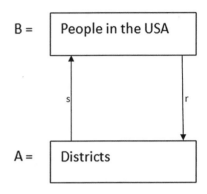

The map *r* signifies the assignment of each person to his or her congressional district, and the map *s* signifies the assignment of a representative of a district, as each representative of a district is a person living in the United States. Lawvere and Schanuel (2000) ask how the map B → B looks where the mapping function is $e = s \circ r$ (i.e., *e* is equal to map *s* following map *r*). Let's assume that John Smith is a citizen of the United States who belongs to the 1st Congressional District of the State of New York. Who is *e*(John Smith)? *r*(John Smith) is the 1st Congressional District of the State of New York and *s*(1st Congressional District of the State of New York) is (as of 2017) a Republican by the name of Lee Zeldin. Therefore *e*(John Smith) is Lee Zeldin, who is the congressman of the citizen John Smith.

One should notice that *e*(John Smith) is not John Smith. However, some people are the fixed point of *e*! These are the elected congresspeople themselves. Lee Zeldin is a fixed point of *e* as he is the representative of the district in which he is a resident.

If *e* is an idempotent map from B to B, then a "splitting" of *e* consists of an object A with two maps *r*(A → B) and *s*(B → A) with $r \circ s = 1_A$ and $s \circ r = e$. The fixed points of *e* are therefore those elements *x* of A satisfying $e(x) = x$. The procedure described above allows us to identify the *number of fixed points* in the category of sets and may be adapted for the identification of centers in wholes.

Imagine that we observe the succession of object A in time where the map *s* signifies a functor from the first appearance (A_1) to the second appearance (A_2) and map *r* is a functor in the opposite direction. We may adopt a softer version of retraction saying that $r \circ s$ is not equal to 1_A but returns us to a point isomorphic to the point from which we departed. In this way and for an object x of A_2, the endomap e establishes a morphism from x to an object formed by $s \circ r$. This object may be roughly considered as a fixed point if $e(x)$ is similar to x up to isomorphism.

We may also think about fixed points not as simple objects (e.g., John Smith) but as *relations or systems of relations preserved under transformations*. For example, when observing a dragonfly in movement, we may see that, regardless of its specific position and posture, its body is always located *in between* its wings. When we map the category of the dragonfly from time slice T_1 to time slice T_2, we may actually map a category in which the objects are the relations between the elements and in which the morphisms are relations between the relations. In this dynamic context, a fixed "point" is actually represented by left adjoint that preserves relations and systems of relations along the transformations of the object.

In sum, the centers of a structure may be identified through the identification of local symmetries (i.e., groupoids) and as fixed points identified through mappings between domains. This process is far from trivial if we are looking for centers of literary pieces (for instance). When examining the characters of a novel, we may intuitively sense that a certain character has a "fixed point," an axis of personality that is quite stable regardless of the character's various appearances in the plot and its developmental trajectory. The way in which local symmetries and centers may be identified in the characters of a novel is an open challenge that deserves a fully developed elaboration.

As mentioned earlier in this chapter, wholeness, as analyzed by Alexander, has the property of being multi-scaled, which means that it can be analyzed as composed of various levels of scale, from components to clusters of components and so on. How do we "know" when we cross a level "up" or "down" scale? The Gestalt slogan already mentioned proposes that a kind of surprise must be associated with such a shift in scales of analysis. We may first identify local symmetries formed around centers. As these local symmetries may vary in terms of their degree of similarity, they can be considered as *alternating repetitions* that form *gradual variations* that lead us from centers to boundaries. The boundaries within each structure are therefore formed around imagined clusters of objects in a way that manifests the synergy of the clustered objects. Boundaries are actually loci where *differences are maximized*. A center is formed between local symmetries. A *boundary* is formed when several objects associating groupoids of various degrees of symmetry constitute a category that is of a synergetic effect, as illustrated in Chap. 3 with regard to the word compound "hotdog." Boundaries form *contrasts* (property no. 9 as proposed by Christopher Alexander (1980/2002)). While this property is evident in visual representations and architecture, it is surprisingly evident in semiotics too (e.g., Assaf, Cohen, Danesi, & Neuman, 2015), where opposition theory suggests that binary oppositions (e.g., good vs. bad) underlie basic cognitive and linguistic processes. The contrast between good and bad, for instance, is formed through negation, which is to recall a reversible process of computation implying two diametrically opposed forms of behavior. A sharp contrast can therefore form a tough boundary, at least in the semiotic context, through negation and the formation of antonyms corresponding with binary values of the value category.

Such basic oppositions may be described as semiotic groupoids as their objects are isomorphic and mutually constitute each other in forming a semiotic binary category. This is of course a common theme both in literature and cinema, where the hero or heroine is often complemented by his or her ultimate opponent: Sherlock Holmes and his alter ego the vicious Professor Moriarty is just one case illustrating the existence of mutually constituting oppositions that cannot be differentiated unless they are mapped to a value category where the mutually constituting objects gain different values that imply different behaviors. There is no cognition without value, and there is no value without behavior.

In sum, the whole is conceived when we first identify local symmetries within an object and between the appearances of the object along a time line. These local symmetries may have variations across different scales in the degree to which the sub-objects constituting these relations are similar (up to the level of isomorphism but not necessarily isomorphic). These building blocks are used to identify centers, which may be considered as fixed points of domain objects identified through their mapping between different domains along the lines presented above. These centers are recursively used to identify local symmetries and their connections. When objects associated with local symmetries (through association in time and space, for instance) cannot be mapped to local symmetries, a connectedness is established with no constraint of similarity. Local symmetries and their relations with other objects are abstracted through similarity, similarity of differences, differences, and differences of similarities at the objects and morphism levels of analysis. When abstracted categories themselves become differences that make a difference, they are considered differentiated clusters of the object up to the level of the whole object, which is different from the sum of its parts as well as different from its surroundings and other objects of relevance.

The emerging whole therefore both is constituted by its elements and constitutes their meaning, as grasped by the concept of the *hermeneutic circle*. This idea may be illustrated in a very simple way. Let's assume that you are reading the sentence, "The cat played the piano." Signs in natural language are polysemic, meaning that they may have multiple senses. The sign Cat, for instance, may literally be used to denote the feline creature, or it may be used in its metaphorical sense to represent a skilled pianist who is a person. How do you know whether you are dealing with a cat in the first or in the second sense?

The whole proposition ("The cat played the piano") is supposed to enable you to make your decision, but how can it support your decision if you don't understand its components? The answer may be quite simple, and it may work well along Peirce's notion of *abductive reasoning*.

As a working assumption, assume that the sign Cat is used in its literal sense. However, when examining in your memory the objects constituting the argument of PLAY[X, PIANO], you may find that most of them fall into the category of a person (e.g., my mother played the piano). Therefore, it would be unreasonable to think that the sign Cat is being used in its literal sense. In practice, things may be much easier as our episodic memory contains situational knowledge that includes particular images.

We may retrieve from our memory representations of events in which we observe someone playing the piano. In most cases, these will be human beings, but we may recall other cases, such as animated movies, in which feline creatures played the piano as well. In order to make a decision about whether the cat in "The cat played the piano" is used in its literal sense or not, we will need to extend the context of our judgment until we reach a point where no more information is gained. We may adopt the heuristic presented before to measure synergy.

As the whole is different from the sum of its parts, we may expect that projections of Cat, Piano, and Play into the whole "The cat played the piano" won't approximate the universal structure of the product of Cat which is taken in its literal sense rather than in its metaphorical sense. We may simply identify the words associated with the context in which Cat, Piano, and Play appear and try to estimate the distribution of this vector by using the vector of the words associated with the meaning of Cat as a feline creature or as a jazz artist. Meaning and wholeness are therefore formed through processes of part–whole relations in a given time slice and whole–whole transformations along a time line.

When judging whether the cat in "The cat played the piano" is the feline creature or a jazz artist, we may also just be patient enough to see what follows the situation of playing the piano. If we keep on reading, the next sentences may improve our understanding. If the "cat" goes to the bar and drinks a shot of bourbon, then it is probably a jazz artist. However, if it curls its tail and orders a glass of milk, then we may suspect it to be the feline creature.

The mutual constitution of whole–part relations can be illustrated with regard to a triangle, which is a simple form, an abstraction corresponding to the "good shape" idea of the Gestalt theorists. The triangle may be considered as a network composed of three vertices and three edges. Each node is connected to exactly two other nodes. When examining the triangle from the *macro*-level perspective, we may ask whether we may identify which node is which based on their macro-level connectivity.[1] The answer is negative. Each node is connected to exactly two other nodes. Therefore, checking the number of nodes to which it is connected isn't informative. What about considering the triangle as three pairs of nodes connected by lines? This move would not help us either. Drilling down from the macro level of the triangle to its components is a move imbued with ultimate uncertainty. What about the other way around? Can we learn about the macro-level characteristics of the triangle? Can they be reasoned out

---

[1] I thank Mario Alemi for this example.

with more certainty based on the micro-level components and their organization at higher scales of analysis?

The answer to this question is negative too. In contrast with artificial structures, which are human abstractions, natural structures are heterogeneous constructs in which "parts" and whole(s) constitute each other. For example, when you see this structure: (Fig. 11.7)

Fig. 11.7 The eye

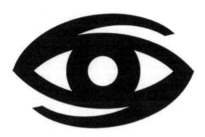

you can infer that another eye exists and that in between you will find a nose and so on. In other words, and in sharp contrast with human abstractions, natural structures (to which natural intelligence adheres) are composed of parts that are informative about the macro-level structure that recursively constitutes the micro-level elements and vice versa.

Summary

- Structures as wholes are formed through various categorical mechanisms.
- Groupoids as local symmetries help us to identify the centers of structures.
- Centers may dynamically form fixed points in a system.
- The hermeneutic circle is evident in structures of natural intelligence.
- In contrast with artificial structures (e.g., the triangle), in structures of natural intelligence, the parts are informative about the whole and vice versa.

Part III

Chapter 12
Let's Talk About Nothing: Numbers and Their Origin

Number is broadly defined as a mathematical object that is essentially used for counting. Indeed, the simplest type of number is "natural numbers," which are used for both counting and ordering and start with zero. We may recall our childhood experience of counting objects. However, as children, we were totally unaware of the fact that the question "What is a number?" is a deep philosophical question. Indeed, performing and understanding counting activities are quite different from understanding the abstract meaning of a number as a "mathematical object" or a symbolic object. This point may be emphasized by realizing that certain animals perform simple counting activities (e.g., Agrillo, Piffer, Bisazza, & Butterworth, 2012; Chittka & Geiger, 1995; Macpherson & Roberts, 2013) despite their lack of a symbolic counting system.

It is apparent, therefore, that we can make number the subject of a careful structuralist analysis of the kind presented so far. Let us start with one philosophical answer to the question "What is a number?"

In his book *Introduction to Mathematical Philosophy*, Bertrand Russell (1919/1993) dedicated a whole chapter to the definition of number. Russell first draws our attention to the fact that a particular number can't be identified with the collection of objects that it signifies. For instance, in Christianity, the number 3 can't be identified with the Father, the Son, and the Holy Spirit. As is well known to semioticians (those who study the general logic of sign systems), the map should not be mistaken for the territory, nor the sign for the signified. Therefore, the number 3, for instance, tells us something about

© Springer International Publishing AG 2017
Y. Neuman, *Mathematical Structures of Natural Intelligence*, Mathematics in Mind,
https://doi.org/10.1007/978-3-319-68246-4_12

all collections of three objects, regardless of the specific types of objects the number signifies as a trio. According to this idea, the number 3 is a sign denoting the class of all classes composed of trios. Put another way, if we group together all collections of two objects, all collections of three objects, and so on, then we get for each collection a "class of classes." For instance, the class "couple" contains all classes with two objects.

It is quite easy, Russell suggests, to find out whether two collections have the same number of objects. Two classes are similar if they have a one-to-one correspondence (i.e., function) between them. This is a function known as *bijection*, which in the context of sets is actually the isomorphism we have encountered before. If we have two sets of objects, then bijection exists between the two sets if they have the same number of objects. Accordingly, the "number of a class is the class of all those classes that are similar to it" (Russell, 1919/1993, p. 18). For example, the number of a trio is the class of all trios, as all classes of trio have a one-to-one correspondence.

In general terms, a number is thus defined by Russell as anything which is the number of some class, and as Russell defined "number" in terms of similarity (i.e., bijection), this definition avoids circularity, so argued Russell. Russell's quite brilliant definition, which has its roots in the work of other mathematicians, is of philosophical interest but doesn't seem to lead us to a better understanding of the way natural numbers are represented by natural intelligence.

According to Russell, number is defined as a class of all those classes similar to it. To comprehend the notion of number, an organism should therefore "understand" that different sets of objects are isomorphic and that this isomorphism can be abstracted to become the class of all classes similar to it. Because he was a philosopher, Russell didn't explain how this process is possible and how it can explain the counting ability evident among humans and some non-human organisms alike.

To try to address this challenge, let's start with the first object of the natural numbers (at least according to some conceptions), which is 0. In set-theoretic constructions of natural numbers, the empty set (denoted as $\{\}$ or $\emptyset$) is the starting point of the number system that is recursively built on the empty set. For example, the first element we have is the empty set, and we apply the recursive function $n + 1 = n \cup \{n\}$ such that:

$$0 = \emptyset$$
$$1 = \{\emptyset\} \text{(i.e., the set that contains the empty set)}$$
$$2 = \{\emptyset, \{\emptyset\}\}$$

and so on.

As we start with the empty set, denoted by 0, it is important to better understand the meaning of zero. Zero is a mysterious mathematical object as it is associated with "nothing." The English etymology of "nothing" explains that this linguistic sign originated in Old English, where it was formed as a combination of negation (*nān*) + thing. "Nothing" therefore primarily assumes the existence of something whose absence is signified through negation, which we should remember is a *reversible* operation from a purely logical perspective, albeit an irreversible and recursive operation from an epistemological perspective. "Zero" also has etymological origins in Sanskrit, where it means "empty space." Zero, the "empty space" that is devoid of a thing, is therefore constituted through "negation," a word signifying in Latin "denial" (i.e., *negatio*) and "saying no." Here a clear link is formed between the etymology of 0 and Freud's illuminating paper "Negation."

Zero as a sign signifying nothing emerged from the concept of negation. Negation may therefore be portrayed in terms of a morphism leading from "something" to "nothing." In other words, given an object, which I signify using the Hebrew letter בּ (Beth), its negation forms the "empty" set that, in its turn, may be used to recursively define the meaning of the object from which it emerged. According to this idea, the primitive notion of zero doesn't have ontological supremacy over any other object but is *relationally* formed through the negation of "something." In the first phase of this process, we may have the map from the object signified as בּ to the empty set:

$$בּ \rightarrow \emptyset$$

Therefore, $\emptyset$ becomes the terminal object of this basic category. Next, we negate the terminal empty object:

$$\emptyset \rightarrow בּ$$

and therefore בּ is defined as a *point*, according to the category theory definition of a point, and $\emptyset$ becomes the *initial object*, which means that it is both an initial and a terminal object, described in category theory as the *zero object*. However, only ב can function as a real terminal object, both as a point and as a meaningful difference, as it is an object rather than its absence. To conclude, the act of negation forms the most primitive notion of zero, which may be symbolized as an empty set or the initial object. As the same logic of negation may be applied to zero itself, a morphism is formed from zero to the object.

Our something – ב – may exist with the same mathematical status as zero, being both initial and terminal, but from the point of view of natural intelligence, something interesting happens, as natural intelligence is always about mapping to some value system. To explain this, let me first introduce the idea of a *well-pointed category*.

A category with a terminal object 1 is well pointed if for every pair of arrows f, g: A → B, such that $f \neq g$ there is an arrow p: 1 → A such that $f \circ p \neq g \circ p$. A well-pointed category is therefore one that has enough points to distinguish non-equal maps. Now, let's consider our basic difference category, where ב functions as the terminal object and Ø sends an arrow f to itself:

$$f : \emptyset \to \emptyset$$

f must necessarily be the identity function as a "nothing" cannot be differentiated further; this is the case for any other form of endomorphism from Ø to itself. However, we may imagine two different pairs of arrows from ב to itself such that each arrow provides us with a different picture of ב within itself. In this case, and if Ø functions as the initial object, then we may construct the category:

$$\emptyset \to ב \to ב$$

where a map p denotes the map from Ø to ב and f and g are two arrows from ב to itself. In this case, we may have a well-pointed category if ב is *self-differentiated* such that $f \circ p \neq g \circ p$. In fact, the well-pointed category above generates the first number – 1 – as we start with the empty set Ø defined through negation, an empty set that recursively defines something by producing two different pictures in ב, one of which is the picture of the empty set (i.e., Ø) and one of which is the set containing the empty set (i.e., {Ø}) signified by the number 1. As you can see, the above theorization explains how numbers may be generated through self-differentiation, and this idea is quite similar to Church numerals which is a way of representing natural numbers through certain higher-order functions.

This idea can be further elaborated if we understand that the differentiation of an object may be recursively conducted in a way that produces a successive list of what may be called "numbers." Playing on the idea of a *natural number object*, we may identify ℶ as such an object that is formed through the negation of Ø and that in its turn functions as the terminal object. The succession of ℶ, previously described as self-differentiation, is therefore a map *s*: ℶ → ℶ, in such a way that any other map from the first object 1 to another object A will form the following structure:

Fig. 12.1 The formation of a number

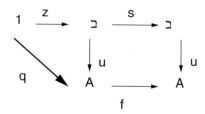

such that $u \circ z = q$ and $u \circ s = f \circ u$, which means that the differentiations may correspond through the succession morphism to what we conceive as natural numbers and their order. In fact, to help us to understand the development of a counting system, Fig. 12.1 should be better explained and extended.

The morphisms among the ℶ objects designate the process of *differentiation*. This process must be accompanied by a corresponding *value* system, which may be represented by the A objects in Fig. 12.1. For instance, a grass eater may differentiate between something – a predator – and nothing. This basic differentiation is crucial as it is accompanied by a critical shift in the value system from a situation of relative safety to one of danger and threat. Let's assume that the "something" is not a single approaching lion but two lions. Does it make a difference?

In terms of the value system, it seems that there is no significant difference between one approaching lion and two approaching lions, as the resulting behavior in both cases is escape. The differentiating process must therefore be associated with a value function that in many natural situations corresponds with a logarithmic or sigmoid function.

There is a third layer that should be added to the above scheme, a layer that uniquely characterizes human beings: the symbolic layer. Since the dawn of civilization, humans have used various tools for counting, from fingers to numbers. Therefore, their ability to differentiate and to load differentiation with value has been upgraded through cultural artifacts that have served as tools for thought.

The three layers of the natural number system (i.e., difference system, value system, and symbol system) may mutually constitute or "lift" each other. For example, the first layer shared both by non-human organisms and human organisms is a differentiation layer, which beyond a certain point has a limited value, as illustrated in the case of the grass eater and the predator. However, the development of a symbolic *naming system* for the representation of these differentiations and their ordinal scale has endowed humans with the ability to count with no theoretical limit in a way that almost naturally invites sophisticated and abstract concepts such as the infinite.

At this point, and through our structural analysis of natural intelligence in general and the (natural) number system and counting in particular, we may better understand the difference between human beings and other "counting" animals. Human beings may possess a more sophisticated differentiation system, specifically if they have artifacts (e.g., sign systems) for representing these ordered differentiations that reduce the cognitive load of the process. At some point, the existence of a symbolic system of representation, indicating the "names" we provide to the ordered differentiations, may gain autonomy from the value system. When a mathematician is conceptualizing the differences between any two successive numbers, she is not considering them as expressing any value system but as existing *in abstraction*. This point can be illustrated with regard to numerosity, or the ability of organisms to represent and compare the cardinality of sets.

In contrast with some models arguing in favor of an abstract representation of numerosity, it seems that non-human, as well as human, organisms use and integrate various sensory cues to estimate numerosity (Gebuis, Kadosh, & Gevers, 2016). If you were asked to judge whether more candies existed in a right-hand jar or a left-hand jar, then you would probably use a heuristic to estimate the numerosity of the candies, for instance, by approximating the relative height and width of the two jars.

In natural situations, heuristics take precedence over careful calculations; thus, as our number system emerged from natural situations, it is reasonable to assume it has been guided by heuristics. When asked to judge which jar contains more candies, your "output" is not an abstract number but a reasonable decision based on a rough approximation where a difference, such as a difference in the size of the jars, is translated into a difference that makes a difference.

The existence of such heuristics is in line with the differentiation and value layers of our model. When asked to judge which jar contains more candies, a difference (in size, for instance) is formed, and this is translated into a difference that makes a difference: we choose the bigger jar, as bigger is sometimes better. The existence of heuristics for numerosity approximation doesn't explain what is unique about the number system developed by human beings, at least not in terms of the underlying cognitive mechanisms responsible for its appearance in certain human societies. However, the idea of numbers as formed through self-differentiation and accompanied by a value and symbolic systems will likely explain the difference that makes a difference between human beings and other counting animals.

In sum, the structure presented above not only may be used to describe the most primitive counting system, as evident even among some non-human organisms, but may also be extended to include the cultural layer, which according to Luria and Vygotsky (1992) is one of the three necessary threads constituting human intelligence. Indeed, it has recently been argued that "relational language" plays a significant role in establishing relational similarity (Christie, Gentner, Call, & Haun, 2016). This conceptualization may help us to try to address some of the current ideas regarding mathematical thinking, as illustrated below. At this point, we may want to deepen our understanding of the number zero.

In a recent review paper, Nieder (2016) suggests that the emergence of zero occurs over four stages. In the first stage, zero means the absence of an object, which is represented as a *neural resting state*. This point is far from clear. The absence of food is clearly an absence experienced by an animal in full emotional modality, and therefore it is difficult to imagine it as grounded in a "neural resting state." Along the same line of reasoning, have you noticed that a newborn moves its lips in a sucking movement in the absence of a breast to feed him?

The breast is clearly absent, but it doesn't seem that the absence is represented in the infant through a neural "resting state." In fact, the absence of the breast is not represented as a resting state but as an *active state*, which is sometimes represented through the infant's fantasies (i.e., imagining it is feeding from a breast). Identifying absence with a "neural resting state" doesn't seem to be in line with the meaning of absence in real-life situations.

In the second stage of the formation of zero, it is argued that the absence is conceived as meaningful but as yet lacking any quantitative sense. This phase is probably equivalent to the vigilance of the grass eater looking out for predators, described above. The existence of nonmeaningful representations is doubtful, though, as beyond philosophical abstractions, there is no reason why natural intelligence should invest energy in a difference that is not translated into a difference that makes a difference.

According to Nieder (2016), in the next stage, the zero is represented as an empty set, and in the fourth stage, the empty set is extended to become the number zero. These stages are not "natural" in the sense that we would only expect to see them among societies in which mathematics has been developed and children are taught mathematical concepts in their modern formal sense. I doubt whether these stages validly represent the way in which zero is represented and evolved in the minds of human beings, and in the next paragraph, I dedicate a few lines to discussing this issue.

The difficulties children exhibit in understanding the meaning of zero and incorporating it into their basic arithmetic raise the question of why is it so difficult to represent the meaning of nothing, which seems to be intuitively identified with a "meaningful" absence. Even if the number zero is embodied, as proposed by some researchers, and grounded in the absence of something, as I've proposed before, it is not quite clear why is it difficult to grasp this abstract symbol of nothingness. This question gains more power when we learn that even animals may learn to represent absence as a behaviorally meaningful category (Nieder, 2016). For instance, in one study, a female monkey learned to use the sign 0 to signify an empty food tray (see Nieder, 2016). In this context, the representation of absence is the representation of a meaningful category that signifies the non-presence of the food the monkey strives to gain. In light of this behavior, one may wonder whether absence has an absolute value with neural correlates (i.e., resting state) or whether absence is a difference category, which is meaningful only as a difference that makes a difference.

To illustrate this point, we may compare two senses of absence. In one context, a grass eater experiences the absence of a predator. The absence, which is of a clear positive value, doesn't seem to correspond with a resting neural state as the grass eater must always be vigilant and actively screen the environment for potential threats. In contrast, the predator observing the grass eater from afar might experience the absence of food. This is clearly a *negatively* loaded experience, and therefore the "same" absence (after all, and from a mathematical point of view, the empty set is one and unique) seems to have two diametrically opposed senses that only when abstracted can be generalized to the concept of zero.

In contrast with mathematical conceptions (which axiomatically assume that zero is unique and the first number), natural intelligence (1) proactively produces absence through negation, (2) proactively produces absence as a working hypothesis that is continuously tested, and (3) generates absence only when that absence is associated with a value system. The difference between existence and nonexistence is therefore basically the same type of difference as that experienced by Köhler's hen, and it is much more complex than the representation of numbers intuitively corresponding to concrete objects and their class bijection.

To gradually develop the abstract concept of zero, a child must therefore identify similarity of differences between different kinds of "absences" formed through the negation of different objects. We are now in a better position to understand why it is so difficult to conceptualize nothingness as zero. From the perspective of natural intelligence, and in contrast with some mystical perspectives, zero is a surrogate of existence and always dependent on the negation of an object. I experience the absence of Food, I sense the absence of a Predator, and I feel the absence of my Mother. In all of these cases, the meaning of absence is dependent on a certain object and its negation. The class of all classes with zero objects is therefore a difficult class to conceptualize as the morphisms are not between objects, which do not exist, but between the arrows symbolizing their negation.

In other words, numbers may be formed through the morphisms between the objects of sets. In the case of zero, the morphisms are established between morphisms of negation that may be associated with different values, and therefore a higher level of abstraction and complexity is involved. Abstraction, as I have explained, is the ability to construct higher-order *relational structures*, which necessarily involves processes of forgetting and compressing. For many years, abstraction has been considered the hallmark of human intelligence. However, as we dwell deeper into the minds of other organisms, we are often surprised to find out how ignorant and dismissive are we. More specifically, a lot of knowledge has been gathered on the ability to abstract among non-human organisms (e.g., Wasserman, 2016). This accumulated knowledge clearly refutes the oversimplified approach to non-human organisms as beasts with no facility for abstract thinking. However, this accumulated knowledge also stresses the qualitative differences between human intelligence and intelligence as it exists among other species. In the next chapter, I will attempt to illustrate the importance of the structuralist approach presented in this book by addressing analogical and metaphorical thinking.

Summary

- Zero as a sign signifying absence emerges from the negation of something.
- Through the formalism of a well-pointed category, we understand how the number system emerges through self-differentiation.
- A number system is coupled with a value system and among human beings additionally with a symbolic system.
- It is very difficult for children to grasp the meaning of zero.
- Natural intelligence among human beings relies on the mutual constitution of the differentiation, value, and symbolic systems.

Chapter 13
King Richard Is a Lion: On Metaphors and Analogies

Analogy and metaphor are two terms used to denote the way in which deep-level comparisons are made between different domains. As the two terms are sometimes used interchangeably and metaphor is considered to be a specific case of analogy and sometimes expressed in terms of X is Y (e.g., "life is a dream"), I will use the term "metaphor" to generally describe a mapping function from a semantic source domain to a semantic target co-domain (e.g., "all the world's a stage") and will use the term "analogy" to describe deep-level similarity of the form A is to B like C is to D (formalized as A:B::C:D), such as Wing is to Bird like Fin is to Fish.

Chapter 7 introduced Köhler's experiment with a hen. We can say that the hen's behavior expressed analogical reasoning in which the relation between the dark gray and the light gray sheets was equated with the relation between the light gray and the white sheets. This finding is surprising as we usually consider non-human organisms to be almost totally governed by perceptual cues (e.g., Smith, Flemming, Boomer, Beran, & Church, 2013). According to this conception, and in comparison with adult human beings, apes as well as young children demonstrate relatively poor performance at tasks of analogical reasoning because they rely on surface structure similarity between the compared items rather than on their deep structure "relational" similarity.

We should remember, though, that the artificial laboratory separation between "surface"-level structure and "deep"-level structure might be misleading, as in real-life situations the two layers are confounded because surface-level aspects may be highly informative about the

Y. Neuman, *Mathematical Structures of Natural Intelligence*, Mathematics in Mind, https://doi.org/10.1007/978-3-319-68246-4_13

deep-level relational structure. For instance, simple perceptual signs of danger are enough to prompt signals of deterrence, which is an abstract relational (deep) structure. We can't ignore the confounding of surface and deep structure relations in real-world situations, and therefore the heuristic value of attuning to surface-level features and similarities cannot be dismissed. This realization doesn't contradict the fact that the superior performance of human beings in executing artificial tasks of analogical reasoning is expressed as the ability to "transcend" surface structure characteristics and to consider deep-level similarities. This ability invites a structuralist explanation of the way analogical reasoning is performed.

It seems that a crucial phase in identifying relational similarity is the development of the ability to detach oneself from the perceptual dimensions of cues and to examine them in abstraction. Let's consider the task of identifying the analogy for:

mason : stone

where the choices are:

teacher : chalk

carpenter : wood

soldier : gun.

photograph : camera

book : word

and the correct answer is:

carpenter : wood

We may first identify possible *relations* between mason and stone. The most immediate relation that comes to mind is "cut" (i.e., mason cuts stone) as masons, or more accurately stone masons, are called "cutters." Therefore, and based on our real-world knowledge encoded in our *semantic memory*, we may produce the hypothesis that the relation associating mason and stone is "cut" and represents this dyadic relation between the two arguments (i.e., mason and stone) in a propositional form, as follows:

$$CUT[MASON,STONE]$$

At this point, we may ask whether one of the other pairs of solutions holds a similar relation. If you go to the Open Information Extraction tool (http://openie.allenai.org) and insert the arguments Mason and Stone, you will find the following relations:

- Laid
- Can make anything in
- Cut
- Would with
- Worked
- Builds in

If you apply the same procedure to the arguments Carpenter and Wood, you will find that this pair shares the relation "cut" with Mason and Stone. Therefore, a reasonable hypothesis is that the Mason cuts Stone just as a Carpenter cuts Wood.

When using the Open Information Extraction tool to identify the potential arguments for Cut, you may find a variety of arguments. Some of these arguments involve the more abstract connotative sense of Cut, such as in the case of Companies that cut Costs. In some cases, the subject is a more concrete object, such as a Chicken that is cut into Strips, and one may find a whole variety of other combinations with either concrete or abstract arguments/relations.

When searching for the arguments of Cut, one may notice the relatively high number of results (298 answers in March 2017). In comparison, the number of relations between Mason and Stone is much more limited (six answers). Therefore, the first step is to identify the possible relations between the arguments of our target pair and to see whether they correspond with relations between the arguments of the potential solutions.

Things may get more complicated if we add a possible solution (such as knife:bread or government:taxes). Indeed, Knife cuts Bread and the Government may cut Taxes, but the solution carpenter:wood is much more appropriate as both Mason and Carpenter are human beings who operate on materials: Stone and Wood. Using the previous heuristic only, our attempt might fail if we add the options of knife:bread or government:taxes.

To address this new challenge, we may simply ask how similar are Mason and Carpenter or Mason and Knife by tracing their inherited *hypernym* in a semantic taxonomy such as WordNet (https://wordnet. princeton.edu), which is structurally equivalent to finding the best semantic object of its type in the universal construction of a co-product where Mason and Carpenter are two objects. In the WordNet taxonomy, the inherited hypernym of Mason is:

mason $\rightarrow$ craftsman $\rightarrow$ skilled worker $\rightarrow$ worker $\rightarrow$ person

In comparison, the inherited hypernym of Carpenter is:

carpenter $\rightarrow$ woodworker $\rightarrow$ craftsman $\rightarrow$ skilled worker $\rightarrow$ worker $\rightarrow$ person

We can see how Mason and Carpenter converge on several of the best objects of their type. In contrast, the inherited hypernym of Knife is:

knife $\rightarrow$ edge tool $\rightarrow$ cutter $\rightarrow$ cutting implement $\rightarrow$ tool

In this case, and despite the fact that Mason and Knife are both arguments that exist in a "cutting relation," Mason is much more similar to a Carpenter than to a Knife, a decision made based on positioning each argument in a co-product structure and examining whether the best object of its type may be found for them. Figure 13.1 illustrates this process of hypothesis testing:

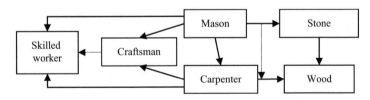

Fig. 13.1 The structure of analogical reasoning

In sum, potential relations between the arguments of the target pair are identified, and we attempt to match these relations with relations characterizing the various possible solutions. In the case of several matching relations, a co-product is constructed on each pair of arguments in order to test the hypothesis that they share the best object of their type.

The similarity–difference thesis is evident in some computational work on the automatic identification of analogies. Indeed, when we look for the best solution for the pair mason:stone, we are looking for a pair that maximizes the similarity between the relations established within each pair and in some cases even *between* the corresponding arguments of the pairs. This is not always the case, as sometimes the arguments share some contextual similarity within each pair but lack any similarity between pairs.

For example, we may form the analogy particle:fluid::bird:sky. This analogy may be interpreted as follows: the Particle moves in Fluid just as the Bird flies in the Sky. In this case, there is a high contextual similarity between the objects *within* each pair (e.g., Bird and Sky) but dissimilarity between the context of the two pairs (i.e., Bird and Particle).

The similarity is therefore expressed as the deep relational similarity between Move and Fly. This similarity is meaningful if we conceive the relations Move and Fly as particular instances of *movement*, a *relation* existing between an object and the medium in which it moves, whether Fluid or Sky. Subsequently, we must somehow abstract Fluid and Sky, and Particle and Bird, by finding the best object of their type. Again, using WordNet, we may find that, through their inherited hypernym, Particle and Bird converge on Physical Entity. This means that, if you abstract Particle and Bird and trace their semantic taxonomy, you will eventually find out that both a Particle and a Bird are instances of a Physical Entity. Similarly, you will find that Fluid and Sky are forms of Matter.

Therefore, we may ask whether particle:fluid actually describes an abstract scenario in which some Physical Entity (the Particle) moves in matter (Fluid) just as another Physical Entity (the Bird) moves or flies in another type of matter (Sky). The idea of the co-product as the best of its type is a powerful tool of *abstraction* that may help us to *optimize* the similarity between a pair of arguments and its analogical solution.

At this point, we may identify different forms of abstraction from those we have discussed so far. The first form of abstraction is that discussed by Piaget and illustrated through the intensional definition of a set. It is simply the idea that a set of objects share some common denominator that is used to group them in the same set. This form of abstraction may be expressed through embodied attributes grounded in sensory–perceptual experience (e.g., all objects having a red color) or through more abstract attributes, such as defining the set of all sets that are not members of themselves.

A second form of abstraction involves the condensation of two objects into a higher-level object that is identified through the structure of the product (or its dual the co-product) as the best of its type. This form of abstraction is evident in the above example, in which a Particle and a Bird are considered as two instances of a higher-level object, which is Physical Entity.

When forming this kind of abstraction, we must discount a lot of information: differences that exist between Particle and Bird, such as that Bird has wings, are thrown away. The price of giving up these differences is justified *only* if the newly formed abstraction (i.e., Physical Entity) generates a difference that makes a difference at a higher level of abstraction, at the level of analogical reasoning. In other words, it is legitimate to discount the differences between Particle and Bird only if the new object Physical Entity contributes new information to the task of trying to identify the analogy between Particle:Fluid and Bird:Sky.

There is another form of abstraction that takes this idea a step forward. When we seek the relation between Bird and Sky, we may decide that the best option is fly_in – i.e., Bird fly_in the Sky. Having identified the relation, we may abstract its arguments by giving them arbitrary names such as B and S. This is actually a way of preserving the *relation only* and representing the arguments as variables – i.e., FLY_IN[B, S]. This procedure heavily relies on the understanding that by identifying a concrete relation (e.g., fly_in), we may give up the concrete arguments (i.e., Bird and Sky) and even their abstraction in a taxonomy and use arbitrary signs as "wild cards." This form of abstraction is quite helpful as we may start identifying different arguments positioned in the dyadic relation fly_in. For example, we may find that Airplane fly_in Sky and therefore that an airplane and a bird have something in common.

Thus far, we have considered *upward* forms of abstraction in which information about and/or between objects is thrown away. However, there is another form of abstraction, which was introduced in Chap. 2 as the *forgetful function*. This function involves a *downward* form of abstraction as its strips out layers of the structure until it reaches the set level, where the objects exist devoid of any higher-order relational structures. For example, if we identify the arguments of fly_in, we may find propositions such as:

1. Bird fly_in Circles
2. Bird fly_in Flocks

3. Goose fly_in Formation
4. Condor fly_in Colca Canyon
5. Flag fly_in Wind
6. Statement fly_in Face of Reason

This set of propositions has a clear relational structure between its arguments. Sometimes this relational structure is literally used (e.g., Bird fly_in Sky), and sometimes it is used metaphorically (e.g., Statement fly_in Face of Reason). We may learn a lot about the fly_in relation by applying a forgetful function and simply grouping the B arguments and the S arguments into two different sets. In other words, we strip down the relation between the arguments and now examine them as objects contained in two sets. Using this procedure combined with other forms of abstraction, we may find out that the source of the metaphorical use of fly_in is of course the embodied activity of observing a flying bird and its extension to other domains. In sum, abstraction can take various forms that when combined may provide natural intelligence with powerful tools of generalization.

Now let's move forward by better understanding what a metaphor is and by analyzing the metaphor "our king Richard is a lion." Let's assume that a child from England is exposed to this metaphor during the reign of King Richard I (1189–1199). He knows that his king is Richard, Richard I, and that the king is a human being. The child may be familiar with the image of a lion, as the lion has been adopted as the sign of the monarchy and appears on army flags, royal regalia, and so on. However, he knows that King Richard is not a real lion.

Therefore, an attempt to map the perceptual image of a lion onto that of the king would result in incongruence. There may be some perceptual similarities between the king and a lion, but the surface structure functor between these categories is quite limited. To recall, human beings have high sensitivity to human faces, and the face of a lion is not very similar to the face of a human being. At this point the child may seek similarities at higher levels of abstraction. To explain this process, we should return to Peirce and his work on relations.

To recall, Peirce (1931–1966, vol, v, 119) conceived any system of meaning to be composed of three basic "molecular" relational structures that cannot be reduced to each other. The basic level of structure consists of an object and its description/predicate, as where a banana is described as yellow. This relation, which is called by Peirce *monadic*, is usually expressed in language through adjectives such as "sweet," "hot," and "green." When a low-level perceptual similarity between King Richard and a lion is refuted, the mind may move forward by listing the monadic relations characterizing each object and seeking a similarity between the two lower-order relations of the objects. As I will show later, this process is not as simple as one might expect, but for now let me focus on this oversimplistic explanation.

The child may recall that lions are described as courageous and that King Richard is described as courageous too. This recollection is based on *semantic memory* and the idea that the child holds some general world knowledge. In fact, the meaning of Lion, signifying someone who is courageous, brave, and strong, first appears in the *Lambeth Homilies* (published in 1175, during Richard I's life), which is an important source for understanding Old English. Somehow the lion came to be associated with courage, and this association is far from trivial. Courage as an emotion first appeared in 1297, and the oldest sense of "brave" appeared in 1546, in the sense of opposing power (i.e., defy). As being courageous and being brave are the same, we may speculate that when European hunters first encountered lions, they were impressed by the way these animals opposed them. The concrete action of opposing may have been transformed into a monadic relation in which lions were conceived as brave, and the next step (cultural abstraction) may then have formed the abstract sign of courage, as will be explained later.

When we process the metaphor King Richard is a Lion, the perceptual aspects of King Richard and the Lion are thrown away, and the similarity of relations is performed on lower-level (monadic) relations first. Given that a perceptual level of similarity has been rejected, the newly formed hypothesis is that King Richard is a Lion in the sense that King Richard is courageous just as a lion is courageous. This metaphor identifies similarity of emotion and attitude as both the king and the lion are emotionally evoked as opposing their opponents. In fact, recall the etymology of courage, which is grounded in the Latin word for "heart," which is the site of emotions. Courage is not a simple and perceptually grounded attribute like color or taste. Finding the similarity between King Richard and the Lion therefore involves an intentional ignorance of perceptual aspects. In this context, we can call to use Peirce's idea of *hypostatic abstraction* (Peirce, 1902) as a process through which a predicate is transformed into an object in a relational structure. For example, the proposition Honey is Sweet may be translated to Honey *possesses* Sweetness. In this example, the embodied adjective "sweet" has turned into an object/argument in a dyadic relational structure composed of two objects (Honey and Sweetness) and one relation (Possess). This process of hypostatic abstraction generates an abstract object, which is "sweetness." While "sweet" is an embodied predicate grounded in our sense of taste and smell, "sweetness" is an *abstract* concept as no one can taste, touch, smell, see, or hear sweetness. When we say "King Richard is a Lion," we may have transformed this proposition into the relational structure "King Richard possesses Lion-ness." While "lion-ness" is a neologism, we can intuitively grasp its sense as equivalent to courage. We will shortly see how this process is involved in the creation and comprehension of metaphors.

The child trying to comprehend the proposition "King Richard is a Lion" may have adopted a simpler process of abstraction by mapping monadic relations from the semantic domain of Lion to the semantic co-domain of King Richard. However, the process of seeking similarities of relations may be conducted at higher levels. For example, when Shakespeare writes in *Romeo and Juliet*, "Death, that hath suck'd the honey of thy breath," he is comparing death to something that sucks honey. Death sucks honey like X sucks honey. Given our world knowledge, we may easily guess that death is here compared to a bee. Just as the bee sucks the honey out of a flower, death sucks honey (life) "of thy breath." This is in fact a triadic relation as it stands for three objects: X sucks Y out of Z. The metaphorical effect is created when we use the relational similarity formed by the above triadic structure and seek relevant objects that may have a correspondence with the above variables:

Death sucks Honey out of Breath :: Bee sucks Honey out of Flowers

The poetic aesthetic effect is formed when (surprisingly) the innocent and positively perceived bee is associated with death, which is negatively loaded. When the bee sucks honey (or more accurately pollen) out of flowers, it does it to nurture young bees and other flowers. This is a life-forming activity. However, when death sucks "honey" out of breath, it is an annihilating force. The dramatic poetic effect formed by Shakespeare stems from the tension embedded in his metaphor, which brings together the two opposing senses of the relations. One sucks honey to support life; the other one sucks life to kill.

Shakespeare's creativity is not expressed through the identification of trivial correspondences between lower-order relations (which are known a priori relations) but in *inventing* and thus forming higher-order correspondences that in themselves are mapped to two different values. In other words, while the mapping from Bee to Death forms a similarity of differences at the object level of analysis, a difference of similarity is formed between the value category of Death and Bee. That is, similarity of differences is formed between Bee and Death through the operation of sucking. However, difference of similarity is established between Bee, which is loaded with positive value, and Death, which is loaded with negative value. It seems that by identifying various forms of similarity of differences, differences of similarities, and their value categories, we may encrypt the meaning of every metaphor or analogy.

Let us return to the metaphor "King Richard is a Lion." The courage of King Richard is explained by mapping certain relations from the more concrete "source" Lion. Indeed, it has been argued by Lakoff and Johnson (2008) that metaphors establish maps from a more concrete and embodied source/domain to a more abstract and vague target/co-domain. In fact, the metaphorical embodied aspect of our cognition has been recognized long time ago by several thinkers, to include Heinz von Foerster, whose work on cognition and epistemology (Foerster, 2003) presented some of the pioneering ideas in cybernetics and constructivist epistemology. Metaphors may therefore be interpreted in functional terms as explanatory tools, but they can also extend the meaning of an original source, add connotations (Neuman, Cohen, & Assaf, 2015), and even be a creative tool for the generation of new senses. In fact, it seems that the core of metaphor is the creative activity (Shanon, 1992), which is grounded in a unique cultural evolutionary context, and that metaphor is not a simple explanatory mechanism as proposed by Lakoff and Johnson. After all, the lion is not a more concrete source than the king, and if one is familiar in advance with the courage characterizing both the king and the lion, what is the point in "explaining" the courage of Richard I by comparing it to a lion?

Even the allegedly straightforward metaphor King Richard is a Lion cannot be comprehended along the lines of Lakoff and Johnson's simple albeit extremely popular theory, but the explanation that I've proposed may lay the ground for a better model. The developmental model that I would like to propose is presented here, and I will introduce it through the metaphor we have already been discussing.

First, we should be aware of the way human beings experience situations and the way these situations are represented as *episodic memory* encoding contextual and emotional aspects of the event. The term episodic memory was first introduced by Tulving (2002) to differentiate between remembrance of personal experience and remembrance of general facts. Knowing that the Eiffel Tower is located in Paris, which is the capital of France, is the expression of general factual knowledge some of us hold in our memory; this is called *semantic memory*. In contrast, recalling one's own visit to the Eiffel Tower is an expression of *episodic memory*. Episodic memory has unique features (Tulving, 2002) as it deals with happenings at particular places at particular times. It deals with "what," "when," and "where."

It is argued by Tulving (2002) that the episodic memory system is unique to human beings and that it allows them to travel back in time and to reexperience past episodes in full awareness. It is argued that the complexity of episodic memories may be accounted for through "scene construction," defined as the process of generating and maintaining a complex and coherent scene or event (Hassabis & Maguire, 2007) and that, therefore, the core meaning of episodic memory doesn't uniquely characterize human beings.

Indeed, it has been argued that proto-episodic memories have a long evolutionary history and that they appear among various non-human organisms as memory for *events in context* (Allen & Fortin, 2013). According to this thesis, there are three behavioral criteria for episodic memory:

1. Content: the individual remembers information about the event ("what") and its context of occurrence (e.g., "where" or "when" it happened).
2. Structure: the information about the event and its context is integrated within a single representation.
3. Flexibility: the memory can be expressed to support adaptive behavior in novel situations.

In the context of human beings, Conway (2009) suggests that episodic memories have nine unique properties:

1. They contain summary records of sensory–perceptual–conceptual–affective processing. This property expresses the fact that episodic memories encode slices of experience, situations, or episodes with their embodied and emotional aspects. It doesn't mean that the memory of an episode is verbatim or necessarily vivid but that it involves strong embodied and emotional dimensions as represented from the unique perspective of the person experiencing the situation.
2. They retain patterns of activation/inhibition over long periods. This property means that, when an episode is encoded in memory, it remains there for a long time as a cluster of objects mutually activating each other. For instance, one may find it difficult to recall a certain piece of information unless one reexperiences the whole episode of which this piece was a part.

The next six properties of episodic memories are self-explanatory:

3. They are often represented in the form of (visual) images.
4. They always have a perspective (field or observer).
5. They represent short time slices of experience.
6. They are represented on a temporal dimension roughly in order of occurrence.
7. They are subject to rapid forgetting.
8. They make autobiographical remembering specific.

The final property is of specific interest for our discussion of metaphors:

9. They are recollectively experienced when accessed. Why do we hold specific memories when general knowledge seems to be enough for survival? A possible explanation is that they serve as cues for gaining general knowledge and for constraining the meaning of general knowledge. For instance, a person who holds a generally "depressive" perspective on life may find it difficult to recall episodes in which he found meaning and pleasure in life. In this case, the general perspective he holds might have detrimental consequences for his well-being. In addition, new knowledge may be acquired by mapping new episodes to old ones and by finding analogies through similarities of differences and differences of similarities. Taking this point further, we may even describe developmental

trajectories through which episodic memories have evolved into general semantic knowledge through which metaphors are formed and comprehended.

Point 9 is one that I would like to elaborate with regard to the King Richard is a Lion metaphor.

First, let's adopt the most general idea of episodic memory: memory for events in context. Second, let's speculate about the ontogenesis of the association between Lion and Courage to understand our metaphor. As mentioned, the hunters who first encountered lions may have been highly impressed with the animals' fierce resistance to their opponents. The encounter would therefore likely have been a fight, or more accurately the lion fighting back against powerful opponents. The most important "what" of this situation is the lion's fierce resistance, which may have made a powerful impression on those involved in the event. As memories tend to decay, what has remained of this situation is the abstract relation of opposing an opponent + the strong emotion. That is, the first phase in the formation of the metaphor may have been the identification of an *abstract relation* (i.e., opposed to) accompanied by a strong emotional arousal and sense of awe.

The next step is the formation of a monadic relation, which is actually a single attribute/predicate attached to our lion as a result of the relation evident in the situation. In the case of the lion, this predicate describes it as *brave* or *courageous*. Please notice that, in contrast with simple predicates grounded in sensory–motor experience (e.g., "sweet"), the predicate "brave" is actually *a single term condensing the abstract relation of fighting with an opponent*. The process of hypostatic abstraction, as proposed by Peirce, allows us to covert the attribute of the monadic relation into an abstract argument, as evident in the following dyadic relation:

$$\text{Possess}[\text{Lion,Courage}]$$

This relation of possession is actually an expression of *the more basic relation of containment*, which may be expressed through the idea of a sub-object. The idea of possession of an abstract property as containment was evident in some past societies where the lion's heart was literally consumed to increase the courage of the hunter, as if the lion's heart, which is contained in the lion, actually contains the lion's courage. The process of metaphorization as described above may be schematized as follows (Fig. 13.2):

Fig. 13.2 The formation
of a metaphor

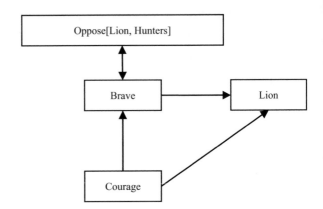

Courage thus becomes a sub-object of Lion as it factors through the attribute, which is actually a way of condensing the information that appears in basic relation evident between Lion and Hunters. Forming the attribute "brave" is therefore a linguistic–conceptual trick used to condense the information that appears in the original situation into a single sign.

In fact, we may apply a forgetful function by saying that anyone who opposes his opponents is brave and conclude that this attribute is contained within him, paradoxically, as an abstract thing – courage. This process involves a mapping from the categories of abstract relations to the category of names/signs corresponding with these relations.

Therefore, when Richard I becomes famous for defying his opponents, one may reason out that he is courageous and brave, like a lion. This metaphor is not functional in the explanatory sense, a point that has been explained before. However, it expresses the way in which the episodic memory of certain individuals, lifted up to the cultural level through language, can be applied to new situations as a natural way of identifying similarities of differences.

The reason why the metaphor King Richard is a Lion is *asymmetric* (i.e., the lion *is not* King Richard) is that the episodic memory of the lion and the process of elevating it to the level of cultural abstract turn the event into an abstract scheme/structure that may flexibly apply to new concrete events. Saying that King Richard is brave is actually the same, in terms of information, as saying that King Richard is a Lion. However, the use of the metaphor may add depth to this description by *reexperiencing* the meaning of being brave through all of its contextual, emotional, and visual dimensions. Saying that King Richard is brave may activate our semantic memory, but using the metaphor

may activate deeper layers associated with images of lions, the powerful emotional valence evoked in these situations, and the vividness of
such reexperienced or reimagined situations.

This process has a deep aesthetic value too. When we say that King
Richard is a Lion, we may automatically evoke the image of a real lion
regardless of the fact that here Lion is used in an abstract metaphorical
sense. As children, we gradually learn the meaning of metaphors based
on concrete experiences. This basic layer of metaphor is still alive in us
as adults, and therefore when we hear the metaphor King Richard is a
Lion, the image of a roaring lion may be evoked in our mind in its full
glory, asymmetrically enriching our imagination of King Richard.

Human beings do not only use metaphors as simple explanatory
mechanisms that map certain deep structure relations from a concrete
and embodied source/domain to a less concrete/embodied target/co-
domain. Human beings don't simply *use* metaphors but *create* metaphors, as is evidenced by metaphors that seem to have an aesthetic value
only. Human beings create metaphors simply because doing so is an
expression of their ability to form and play with abstract structures,
combining them with powerful symbolic and value systems. When
Shakespeare writes in *The Tempest* that "we are such stuff as dreams are
made on," he compares human beings to dreams. However, one cannot
trivially identify the correspondence between human beings and dreams
without opening up a whole host of interpretations.

One possible interpretation is that dreams are temporary and illusory
states. Human beings are clearly not illusory states as they have a concrete form of existence. However, when Shakespeare draws the metaphor, he doesn't necessarily assume that "illusory" is a known attribute
of human beings but *creates this new attribute through the metaphor*!
That is, the mapping function used by Shakespeare forms in the category of human beings a new predicate, which is their illusory existence.
This metaphor allows human beings to reflect on their life in a new
light. Regardless of our concrete existence as immersed in matter, specific perspectives, unique individuality, pain, and happiness, we may all
be illusory, just like a dream. This creative metaphor may stem from a
symmetric reversal of rules evident in other philosophical reflections as
well; when one dreams a dream, the dream is contained, so to speak, in
one's mind. We are the dreamers of our dreams, but what if we are the
dreams of our dreams? This sounds like a ridiculous question, but in its
general form, it is evident in other cases as well, and its deep psychological aspects will be discussed in the next chapter.

The Chinese philosopher Zhuangzi described a situation in which he dreamed he was a butterfly but then awoke to wonder whether he was a man dreaming he was a butterfly or a butterfly now dreaming he was a man. There are close links between metaphors, their deep structural aspects, and such wild philosophical speculations involving symmetric relations between asymmetric objects. This idea will be elaborated in Chap. 14, where I discuss deep mathematical structures of the unconscious. However, let us return to the English child who is trying to process the metaphor "King Richard is a Lion."

When the child first heard that King Richard is a Lion, he may have imagined the king as a lion with a crown. However, as he grew up and acquired both a better ability to abstract and the abstract schemes of his community, he would have come to perfectly understand that King Richard is not a lion in the sense of the feline creature that he has never observed in real life (except for in visual images) but that King Richard is brave, just like other figures, human or non-human, who defy their opponents with great passion. When reasoning out the exact meaning of the metaphor, the child couldn't articulate the exact way in which the sense of courage has been attached to the lion through a long cultural evolution but then expressed in his own understanding, an understanding that echoes deep layers of natural intelligence. We probably share the same basic mathematical structures underlying intelligence with other non-human organisms. However, while Köhler's hen may share with the English child the same logic of similarity of differences, it is doubtful whether the hen can reflect upon its self as "stuff as dreams are made on."

Summary

- Human beings can apply analogical reasoning to symbolic objects, not only to categories immersed in the perceptual realm.
- Analogical reasoning may involve various mechanisms of abstraction, such as abstraction through the product/co-product.
- Metaphors may be culturally grounded in episodic memory that has been elevated and translated into semantic memory.
- The mechanism of hypostatic abstraction may be of great value in understanding the formation of metaphors.
- Metaphors cannot be explained through a simple functional perspective, as they are acts of creativity resulting from the general mechanisms of natural intelligence.

Chapter 14
The Madman and the Dentist: The Unconscious Revealed

The idea that our mental lives are mostly unconscious – that is, beyond our reflection – has appeared throughout history but ultimately appeared as a powerful explanatory theoretical construct in Freud's psychoanalysis. There is an in-built difficulty in trying to use the unconscious as a theoretical explanatory construct. As the unconscious signifies the realm that is beyond our reflection, we cannot have direct access to it.

For instance, let's assume that the legendary Robin Hood is being arrested for attacking the king's tax men. When interrogated by the Sheriff of Nottingham about the reasons for his rebellious act, he may testify, from the *first-person perspective*, that he was motivated by a moral interest to remove the burden of tax from the poor. That is, when asked to reflect on his motives, Hood can easily provide the reasons for conducting such an act.

A psychologist may question Hood's explanation and his basic ability to reflect on his *real* motives. Some of our motives are clear to us but some are not. The psychologist may offer another explanation, which is that rebelling against the king is unconsciously rebelling against an *authority figure*. The problem here is that, in contrast with the reflective and first-person-perspective explanation, it is quite difficult to "ground" the explanation and validate it. We may ask the psychologist: How do you know? Furthermore, in contrast with Hood's answer from "the horse's mouth," the explanation from the unconscious has no anchor in the person's own report.

© Springer International Publishing AG 2017
Y. Neuman, *Mathematical Structures of Natural Intelligence*, Mathematics in Mind,
https://doi.org/10.1007/978-3-319-68246-4_14

The major contribution of Freud was not in generally introducing the idea of the unconscious but in trying to *encrypt the logic of the unconscious, which is a unique logic of translation*. This point is illustrated when we try to unencrypt the meaning of dreams, as dreams are not simple recollections of events but involve a unique logic of translation, as expressed, for instance, in *condensation* (where two separated objects are merged into one), and so on. In fact, Freud describes the logic of the unconscious as a kind of a mirror world of the rational realm of logical thinking. This "mirror world" is clearly evident in literary pieces from *Alice Adventures in Wonderland* (Carroll, 1865/2009) to *Hard-Boiled Wonderland and the End of the World* (Murakami, 1985/1993). It seems therefore that the core of the Freudian venture lies in trying to elucidate the deep "mathematical" logic of the unconscious, which may be generally conceived as the mirror world of logical rational thinking.

It must be noted that the ideas of "logical" and "rational" thinking should not be simply taken for granted, specifically in the era of "reflective modernism." However, there seem to be basic principles of rational thinking that, if given up, would pull out the carpet underlying modern science and what we conceive as normal adaptive thinking. For instance, consider the *law of identity*. As I suggested in Chap. 2, a person who doesn't know who he is and lacks a basic sense of self-integrity should be suspected of a deep personality problem. A science that believes that the law of identity doesn't hold and that everything is in flux and therefore nothing can be said cannot advance beyond mere philosophical speculation. Therefore, and regardless of our reflective and critical stance, we must accept the basic logic of rational thinking and appreciate Freud's attempt to formulate the logic of a rational thinking "mirror world."

At this point, it is also important to realize that the two "logics" are not mutually exclusive and that they are deeply interconnected with each other and mutually *nurture each other*. There is a deep reason why the two systems mutually nurture each other and I will briefly explain it. The realm of the conscious, as will be explained shortly, is the realm of *asymmetric relations*, in contrast with the realm of the unconscious, which is governed by *symmetric relations*. An asymmetric relation forms a clear boundary, such as in the case that I'm the father of my children but my children are not my parents. A clear line or a boundary is drawn between me and my children. It is necessary to form boundaries as otherwise thinking would have diffused in such a way that adaptation would have been impossible. In nature the ultimate symmetry means death ....

Transcending the boundaries formed by asymmetric relations requires turning them into symmetric relations. This is a hallmark of human creativity. For example, I'm the author of this book and I'm currently writing about myself. From a rational asymmetric stance, I'm the author of the character currently described in the book. Can this relation be reversed? In the novel *Hide and Seek* by Dennis Potter (1990), the hero is tortured by the idea that he is a character in a book. This creative idea is possible only by applying symmetric relations.

In sum, rational and irrational thinking should not be simply conceived as opposing and mutually exclusive. I'm always saying that the ultimate rationalist must necessarily be a mystic as acknowledging the boundary of rational thinking implies the acknowledgment of what is beyond, and vice versa.

The deep structural aspect of Freud's treatment of the unconscious has been seriously considered by Ignacio Matte-Blanco (1998), a psychoanalyst who sought to reformulate the Freudian unconscious in logical or mathematical terms. Matte-Blanco's (abbreviated from now on as MB) basic thesis is that the unconscious should be understood as a symmetric form of logic. His *principle of symmetry* is actually a form of mapping where an asymmetric relation is transformed into a symmetric relation. It is clear that this principle of symmetrization destroys many types of binary differences whether spatial (up and down), temporal (before and after), or social (being the father of ...).

This principle of symmetry is also applied to the relation of *part–whole*, where an element of a set is used to signify the whole set and vice versa. This symmetry is evident in the case of *synecdoche*, a rhetorical tool in which an element of the whole is used to signify the whole. One example is the use of "boots" as a shorthand for "soldiers." The symmetrization also exists between elements of the same set. For example, Robin Hood might have considered the Sheriff of Nottingham's treatment of him as "humiliating." In his mental set of humiliating things, he may have found both his domineering father and the Sheriff, and therefore the Sheriff may have been equated with his father. This is actually the logic underlying the metaphor that analyzed in the previous chapter (King Richard is a Lion). As both the king and the lion are elements of the set of brave objects, they can be substituted through symmetrization.

This *process of generalization* is illustrated by MB with regard to one of his schizophrenic patients, who told him, "Your assistant is very rich" (1998, p. 45). When he asked for an explanation, she replied, "He is very tall." Implying that the assistant is rich given that he is tall implies the abstraction of these attributes to a higher class characterized as "high." If someone is tall, it means that he is high in the physical sense. If someone is rich, he may be conceived to be a member of the high economic class. However, symmetrization through which tall and rich are equated has no logical or empirical basis.

Another nice example of the logic of symmetrization given by MB is a schizophrenic patient who went to the dentist after being bitten by a dog. The patient's behavior needs explanation, as a dentist is not usually a person's first choice for treating a bite. There is a possible and simple explanation for this choice. The dentist is a "doctor," and for the patient anyone who is called "doctor" is a physician who may treat a wound. I've observed this mistake among uneducated people from traditional societies. However, MB seeks an explanation of the patient's behavior in different and much more complex terms. Whether his explanation is more valid than the simple explanation that I've proposed is an open question, but at least MB's explanation illustrates his thesis.

MB suggests that if the dog has bitten the patient, then the patient has symmetrically bitten the dog. As the dog is "bad," so is the patient. As the patient is bad, his parts are bad too. As the teeth are a part of the patient, they are bad too. Bad teeth should be taken care of and who is more competent at taking care of "bad" teeth than a dentist? The conclusion of visiting a dentist to get treatment for a bad tooth seems to be perfectly rational, but in the above context, it expresses alternating asymmetric/symmetric relations – what MB calls "alassi" (from the phrase "alternating asymmetrical symmetrical structure"). One must admit that this is quite a creative explanation, regardless of its validity.

The processes of symmetrization introduced by MB are clearly evident in mystical poems. For instance, a mystical Persian poet by the name of Jami (1414–1492) describes humans in one of his poems (translated by W. C. Chittick (1981)) as a reflection of the "eternal light" and the world as a wave of an everlasting sea. If light is reflected in a man, like an object is reflected in a mirror, then the man contains this eternal light just as a mirror contains its reflected image. Substituting the object "man" for his sub-object "eternal light," we may understand that man *is* eternal light. This idea echoes the ancient idea of humanity as created in God's image (*imago Dei*). The *eternal* light may be symmetrized with the *everlasting* sea as "eternal" and "everlasting" are synonyms. Therefore, Man *is* Light and Man *is* World, and there is no separateness between Man, World, and God as all of them reflect each other. This is of course a clear lesson of mystical thinking pointing to the deep interconnectedness of existence.

The same logic may be used to interpret another poem by another very famous Persian poet: Hafiz of Shiraz (b. 1325). In one of his poems, "The Green Sea of Heaven" (tr. Elizabeth T. Gray Jr., 1995), he describes himself as the slave of love but also as a "bird" from heaven and "angel" from paradise who fall to the earthly world as a result of Adam's sin. The poet describes himself as a slave waiting at the door of "love's tavern." The freedom of the angel and the bird in paradise has turned into the slavery of love. If he is a slave, then who is the master? Love is one master, but the poet also appeals to God as his master and lord. An exchange is therefore formed between God and Love: Love *is* God. This metaphorical move ends with the OR operator, where the poet calls for his loved one to wipe the tears from his face with her soft curls OR he will be uprooted just as he has been uprooted from heaven.

MB presents an interesting explanation of why our symmetric thinking is unconscious. Symmetrization produces an enormous number of relations between mental objects; it is an expanding activity, a vast network of connections that can be represented only as a high-dimensional structure (MB, 1998, p. 91). MB argues that trying to consciously process this vast network on the table of our working memory would result in an impossible cognitive load, just as we might experience (for instance) when trying to represent and follow the computations of a deep neural network. Conscious thought is therefore simpler, argues MB, as it is of lower dimensionality and allows us to reflect on quite simple processes, such as that if A entails B and we observe A, then we may conclude B.

However, MB adds an interesting layer to the way the unconscious is expressed in conscious life, which is through repetition. He gives us a simple example. Imagine a triangle composed of three nodes: A, B, and C (Fig. 14.1):

Fig. 14.1 A triangle

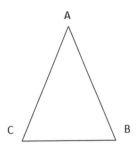

Projecting this structure to a lower dimension, we will have:

$$A \rightarrow B \rightarrow C \rightarrow A$$

That is, we will have a repetition of A. A repetition, says MB, is an indication that a higher-dimensional structure has been projected downward onto a lower-dimensional structure. It must be noted that repetition may involve the repetition of objects of various forms and modalities. The repetition may be of a sign in natural language or of its semantically isomorphic forms. It can be of a visual icon, of a sound, or of *anagrams* (Sasso, 2016). The specific aspects of the repeated object are less important than the dynamics of the repetition.

The idea of a repetition as signaling a projection onto a lower dimensionality is interesting, as repetitions have also been mentioned by Christopher Alexander (1980/2002) as one of the basic properties of wholeness, and here they are used as an indication of a projection from a higher-dimensional and symmetric structure of the unconscious to the lower and asymmetric realm. Indeed, the path from the first appearance of A to its second appearance demarcates the boundary of the triangle, and therefore, the composition of morphisms forming a cycle may be an indication of a structure.

In addition, the repetition of an object may also be indicative of a more general sense of a fixed point. An endomap applied to a mental domain would not necessarily result in a fixed point as the outcome of each operation. However, a path of compositions visiting the same object may be considered in a looser sense as pointing to an attractor that is a zone of interest and preoccupation. From a psychological dynamic perspective, this idea is evident in *fixation*, where a subject is obsessively occupied with a single object. A person obsessively occupied with relations of power may not express identity of objects in the strict mathematical sense. Such a person may not identify his boss with his subordinating father but may definitely establish a similarity or equivalence between these two figures up to isomorphism. In this way, our neo-structuralist analysis may enrich MB's conception of the unconscious by using relations as objects, structures at higher levels of abstraction, and category theory structures to better understand the basic ideas underlying the logic of the unconscious.

Let's use the above theorization in order to understand one phrase, which is "to eat one's hat." This is a metaphorical phrase that may also reflect unconscious layers. This phrase states one's readiness to eat one's hat if an event one is certain will not occur does in fact surprisingly occur. But why should one express one's readiness to eat a hat as a sign of one's confidence that something won't occur?

The meaning of eating is basically to consume food. However, in English, it has an additional old meaning, which is to destroy, kill, and torment, as expressed in the Old Saxon verb *qualm*, which means "to kill." Eating also has a cannibalistic sense as it appears in the sense of devouring a person. The titan Cronus devouring his son in Goya's picture *Saturn Devouring His Son* is just one example of eating as cannibalistic destruction.

People who consume food have probably realized that consuming food involves the destruction of its original structure. Eating is not a structure-preserving transformation but a structure-*destroying* transformation. Whether one is chewing an apple or a piece of meat, the original form of the food is being destroyed. The relation formed through eating, and as expressed in the transitive verb "eat," involves the asymmetric relation between the eater and the food eaten. If X eats Y and eating is a destructive activity, then eating and destruction may be equated. Moreover, if the eater eats/destroy the food, then through symmetrization, the eater may turn into food, and the food might destroy him or her. Promising to eat my hat as a sign of confidence is therefore making an aggressive statement that involves the risk of acting *destructively* against my *own* self. But what does the hat have to do with it?

The *Oxford English Dictionary* mentions an older form of the phrase, which is "I'll eat (Old Rowley's) hat." "Old Rowley" was Charles II's favorite horse and also a name used for Charles himself, supposedly in association with his sexual reputation. After all, the word "stallion" exists in the sense of a male of intensive sexual urges, a "horeling," from "whore" + "ling," which is a suffix indicating a person associated with … whores.

So, a king is riding a horse. He is associated with the horse. The person has the reputation of a "horeling," and therefore, he is a "stallion" and is given the name of his horse. Eating the hat of the "human stallion" is an expression of synecdoche as the hat is used in the sense of a social position. Eating the stallion's hat is therefore an indirect threat to destroy/eat the king himself, which symmetrically might destroy the eater of the hat. Now, threatening a man who has great symbolic and sexual power is a real risk, and someone who is ready to take this risk probably assumes that the probability it will happen is zero.

Similarly to the way the King Richard metaphor has emerged, the specific episode of King Charles and his reputation as a womanizer may have unconsciously and creatively formed the phrase "to eat my hat." MB's formalization of the unconscious as involving processes of symmetrization, coupled with the structuralist approach presented so far, may provide us with powerful analytical tools for understanding various expressions of human creativity, to include the abovementioned phrase.

Summary

- The unconscious was studied by Freud from a structuralist perspective.
- It has been mathematicized by Matte-Blanco, who argues that processes of symmetrization govern the unconscious.
- Analyzing processes through which creative structures are formed may bring to light unconscious symbolic processes.
- This proposition is illustrated through an analysis of the famous phrase "to eat my hat."
- Etymological analysis combined with a careful structural analysis may shed new light on creative processes as expressed in human symbolic systems.

Chapter 15
Discussion

In the Preface, I objected to the notion of "mastering" wisdom and adopted the metaphor of wisdom as rivers. While the Jewish Talmudists who proposed this metaphor were probably not aware of Heraclitus' *panta rhei* ("everything flows") and considered human wisdom only, the two metaphors can be merged even in the context of nonhuman organisms. Living as a part of the natural world requires a continuous process of adaptation to the everlasting flow. For this reason, the idea of having full mastery of "general truths" is an empty pretension both for the human being and the virus and can be warmly adopted only by the *Homo academicus*, who might be arrogant enough to blind himself to the existence of reality.

Having said that, I do not support any simple form of naive realism or a relativist postmodernism, two approaches that should be firmly rejected. Both the natural and the cultural–social worlds exhibit various forms of order. However, some of these orders must be continuously attuned to as a result of the dynamic nature of the world. The virus perfectly understands that it must adapt. A virus that holds a naive realist perspective will imagine a simple ordered world and therefore will never adapt. Its twin virus, the postmodernist virus, may try to shape its *own* structure as what is important is not the world but the "narratives" we build of the world. This kind of virus will not survive either.

The challenge facing all living creatures (of every level of complexity) is to develop dynamic models of real-world patterns. This is what natural intelligence is all about. The old definition of intelligence

© Springer International Publishing AG 2017
Y. Neuman, *Mathematical Structures of Natural Intelligence*, Mathematics in Mind,
https://doi.org/10.1007/978-3-319-68246-4_15

as the ability to know "general truths," here interpreted as patterns, holds both for the virus and the human being.

I have emphasized the fact that, in itself, the venture of artificial intelligence, notwithstanding its impressive achievements, cannot be used as a template to model natural intelligence, for the same reason that the digital computer cannot be used to model the human brain. Artificial intelligence, as currently epitomized in the deep learning approach, is mainly about the development of "intelligent" tools. Indeed, the reflective leaders of this field emphasize that they have no pretensions to model the brain or human intelligence in general.

It then follows that, where living systems are "patterned" to a dynamic world they must model, asking questions about the modeling process is an inevitable scientific move. The social sciences, psychology among them, are interested in identifying variables and in modeling social and psychological phenomena by studying the relations between those variables. The structuralists offered a different alternative, which is to identify general structures for modeling human activity and products. In this book, I have attempted to adopt the same structuralist perspective, but this time with an approach that corresponds with some open questions and past failures.

I have considered a structure as a category, in the mathematical sense, and rejected the simple and naive notion of structure. I found the categorical interpretation of structure to be appealing, as categories may be defined in terms where relations have precedence over objects. This rather counterintuitive idea may serve us when modeling a dynamic world. If one has to adapt and understand a dynamic and noisy world, then one should first model *relations* rather than objects. However, even this idea can be criticized based on the understanding that relations themselves may be the subjects of change. This is of course true. However, if we imagine natural intelligence as a "machine," a metaphor that is illuminating as a metaphor only, then it is a very unique kind of machine. It is not a machine for the processing of a priori and stably defined relations but a meta-machine for generating relations that process machines. While this idea may sound obscure, it has an almost perfect analogy in the adaptive immune system (Cohen, 2000).

The mathematical structures I've used in this book should not be interpreted as *the* structure but as Lego-like blocks used to model the various structures we encounter in studying natural intelligence. This is the context in which I introduced the groupoid as a building block of structure.

The groupoid has been interpreted as a local symmetry dynamically formed through the "synchronization" of several objects. The use of the groupoid as a building block of structure is justified as it maintains the Piagetian holy trinity of a structure: wholeness, transformations, and self-regulation. The groupoid's reversibility is assured as its isomorphism is interpreted in dynamic terms of synchronization. However, as the building block of a structure, the groupoid is meaningless unless it is contextualized in a system in which some information is inevitably lost as a byproduct of irreversible processes of computation. I have illustrated this idea in the context of neural circuits, using the logic of the groupoid to model local, recurrent, and excitatory circuits in the cortex. I have also argued that the groupoid structure explains the integration of neural information flowing upward before some of it is lost as a result of an irreversible process of computation.

Identifying the meaning of groupoids, or local symmetries, in the wild is far from easy. I've pointed to the fact that time is a crucial dimension for experiencing objects in the real world and that, when we follow objects in real situations, we are actually guessing that the successive appearances of an object in time may be considered as appearances of an isomorphic object. Hence, the appearances of an object in time form a groupoid. Moreover, gaining an understanding of what certain objects are involves not only the inference of their structure-preserving transformations but also the identification of higher-order mappings between the transformations of an object in time. Reflective abstraction, according to Piaget, is the abstraction of transformations, and this process may be rigorously studied through the idea of the functor (a map between categories) and through functors of functors.

Acknowledging the limits of a pure mathematical formalism, I have proposed that the strong notions of identity and isomorphism should be released in favor of category equivalence. This move doesn't mean the release of mathematical rigorousness, as category equivalence may be rigorously defined.

This idea opens up the possibility of identifying local symmetries and understanding other interesting mental processes, such as analogical reasoning. Indeed, Köhler's experiment shows us that even a hen can apply a certain form of analogical reasoning by being attuned to the deep relational structures of its domain. Köhler's experiment also illustrates the fact that our natural intelligence is deeply embedded in a value system, which means that all living systems are involved in what elsewhere I have described as "meaning making" (Neuman, 2008).

There is a long scholarly tradition that draws a sharp boundary between, on the one hand, the *res cogitans* and pure "godly" reason and, on the other, the *res extensa* with its earthly and "despised" corporeality; however, it seems that this division, which is manifested today in the division between information and meaning, doesn't hold for living systems. From the amoeba to Shakespeare, we all process differences, similarities, differences of similarities, and similarities of differences. We all process these relations only if they are meaningful and may be mapped to a value category.

The difference between Shakespeare and the amoeba is not so much in the general principles but in their level of complexity and correspondence with value and symbolic systems. While the amoeba may process simple relations, Shakespeare could process and creatively generate analogies and metaphors. While for the amoeba the only value system is a simple reward system motivating a straightforward form of *reinforcement learning*, for Shakespeare, the value system may also include aesthetic values built on layers upon layers of cultural development. For the amoeba, the symbolic system may involve simple indexicality, as expressed (for instance) in signs pointing to a rewarding source of food. For Shakespeare, the symbolic system involved a rich system of signs in which words may be polysemous, invented, and deeply related to their context through network circuits (Danesi, 2003).

All living systems react to a "mediated" reality, whether they have a nervous system or not. Human beings are of a unique status, as their way of representing the world, whether the external, internal, physical, cultural, or psychological, has been developed to an unprecedented level of sophistication. Thus, the structures underlying human intelligence cannot be studied in the same way as the structures of other organisms or in the same way as the structures of natural intelligence that characterized our ancient ancestors.

The modeling of the modeling processes is therefore a complex issue, and one aspect of this complexity has been discussed under the title of uncertainty. This term, which is associated with chance, has almost been given a single dominant sense in the modern social sciences, which is the one proposed by probability theory. Being ignorant of this interpretation and cultural–historical context might turn us into fundamentalist thinkers who consider concepts to be reified and sacred entities. However, the probability theory interpretation of uncertainty and chance is only one possible interpretation among many. Similarly, mathematicians dealing with the realm of ultimate abstractions long ago realized that even Euclidean geometry may have an alternative in the form of non-Euclidean geometries, and social scientists dealing with the overwhelming complexity of the human mind in its various forms of expression and organization must not be fixated on a single form of uncertainty invented in the context of games of chance. God doesn't play the dice, as argued by Einstein in a totally different context, and nor do living systems.

In this context, and as a part of the neo-structuralist perspective presented in this book, I've proposed a novel conception of uncertainty. This new conception doesn't aim to dismiss the one proposed by probability theory, just as non-Euclidean geometry doesn't aim to dismiss or replace Euclidean geometry. The idea I've proposed suggests that, as natural intelligence deals with forming maps of the world, as formalized by our category theory structures, uncertainty may be expressed through the *incongruence* of maps. I called this the *natural transformation modeling principle*, and I proposed that this principle may explain how we gain knowledge *ex negativo*, from our ignorance. The incongruence of our maps may indicate that we observe variations in the same pattern over time, in part–whole relations of the same pattern, or in different patterns. These three basic hypotheses of structural uncertainty may be highly informative, and they don't require any prior knowledge or assumptions regarding the true nature of the "distributions" we observe.

As for part–whole relations, I've dedicated a lot of space to discussing the "wholeness" aspect of a structure by trying to explain it through the rigorous mathematical lines presented in the book. While

I've linked my arguments with Christopher Alexander's seminal work (which, surprisingly, I only came to be familiar with after writing my main thesis), my formulation of wholeness aims at a higher level of mathematical sophistication than the one evident in Alexander's book.

Furthermore, I've suggested that groupoids, as local symmetries, may help us to identify the centers of structures, centers that may be conceived from a structuralist–dynamic perspective as fixed points of the structure-in-transformation. Establishing local symmetries and fixed points at various scale of analysis may allow us to understand the recursive–hierarchical structure of patterns identified by natural intelligence, where the whole supports the parts that recursively support the meaning of the whole. This *hermeneutic circle* seems to characterize various patterns, as identified by natural intelligence in fields ranging from percepts to texts.

At this point, the major thesis of the book, its potential benefits, and its limits and pitfalls may be evident to the reader. The book's main benefit may be summarized as a challenging experience for the intellectual. I will be satisfied even if this limited benefit comes to be acknowledged. However, there are three additional potential benefits.

The first benefit is in better understanding the way natural intelligence identifies and forms orders existing in the world. I've tried my best to show how the neo-structuralist perspective provided in this book may help us to make sense of a variety of phenomena in fields ranging from neuroscience to natural language.

The second potential benefit of this book is therefore in providing theorization, which is justified on the theoretical level by providing "knowledgeable descriptions" of natural intelligence's various expressions.

The third potential benefit of the book is a practical one. The reader may find the theoretical Lego blocks provided herein relevant for modeling processes in his or her own domain, for addressing real-world challenges, and for developing practical solutions to those challenges. For instance, researchers interested in designing intelligent systems may find the ideas presented in this book to be highly relevant, even if only as inspiration.

Having discussed the human ability to form metaphors, I can't avoid concluding this book with a reflection expressed in metaphorical terms.

Writing an academic book is like sending a message in a bottle. One can never know where it will end up and what its impact will be. For me, the personal justification for sending such a message in a bottle can be expressed in terms of another metaphor, which is the Talmudic metaphor of wisdom as rivers that cannot fill one's heart (the sea). Our wisdom can be enriched only by pouring it into other vessels (hearts) with which we are entangled. Shakespeare may have got it right when he suggested that we are such stuff as dreams are made on, and he would probably have accepted a version of the Talmudic metaphor, which may say that we are bottomless vessels of dreams that can never be filled but only enriched, paradoxically by pouring wisdom into others.

References

Adámek, J., Herrlich, H., & Strecker, G. E. (1990). *Abstract and concrete categories*. Chichester, UK: Wiley.

Agrillo, C., Piffer, L., Bisazza, A., & Butterworth, B. (2012). Evidence for two numerical systems that are similar in humans and guppies. *PloS One, 7*(2), e31923.

Alexander, C. (1980/2002). *The nature of order: The process of creating life*. Berkeley, CA: Center for Environmental Structure.

Allen, T. A., & Fortin, N. J. (2013). The evolution of episodic memory. *Proceedings of the National Academy of Sciences, 110*(Suppl. 2), 10379–10386.

Amir, Y., Harel, M., & Malach, R. (1993). Cortical hierarchy reflected in the organization of intrinsic connections in macaque monkey visual cortex. *Journal of Comparative Neurology, 334*(1), 19–46.

Assaf, D., Cohen, Y., Danesi, M., & Neuman, Y. (2015). Opposition theory and computational semiotics. *Σημειωτκή: Sign Systems Studies, 43*(2–3), 159–172.

Attanasi, A., Cavagna, A., Del Castello, L., Giardina, I., Grigera, T. S., Jelić, A., ... Viale, M. (2014). Information transfer and behavioural inertia in starling flocks. *Nature Physics, 10*(9), 691–696.

Bakhtin, M. M. (1990). In M. Holquist & V. Liapunov (Eds.), *Art and answerability*. Austin, TX: University of Texas Press.

Bateson, G. (1979/2000). *Steps to an ecology of mind*. Chicago, IL: University of Chicago Press.

Bateson, G., & Bateson, M. (1988). *Angels fears: Toward an epistemology of the sacred*. New York, NY: Bantam Books.

Battiti, R., Brunato, M., & Mascia, F. (2008). *Reactive search and intelligent optimization* (vol. 45). New York, NY: Springer.

Bennett, C. H., & Landauer, R. (1985). The fundamental physical limits of computation. *Scientific American, 253*(1), 48–56.

Bohm, D. (1998). In L. Nichol (Ed.), *On creativity*. New York, NY: Routledge.

Carroll, L. (1865/2009). In H. Haughton (Ed.), *Alice's adventures in wonderland and through the looking-glass and what alice found there*. Harmondsworth, UK: Penguin.

Chittick, W. C. (1981). Jami on divine live and the image of wine. *Studies in Mystical Literature, 1–3*, 193–209.

Chittka, L., & Geiger, K. (1995). Can honey bees count landmarks? *Animal Behaviour, 49*(1), 159–164.

Christie, S., Gentner, D., Call, J., & Haun, D. B. M. (2016). Sensitivity to relational similarity and object similarity in apes and children. *Current Biology, 26*(4), 531–535.

© Springer International Publishing AG 2017
Y. Neuman, *Mathematical Structures of Natural Intelligence*, Mathematics in Mind,
https://doi.org/10.1007/978-3-319-68246-4

Cohen, I. R. (2000). *Tending Adam's garden*. New York, NY: Academic.

Cohen, I. R. (2017). Running with the wolf of entropy. Manuscript in preparation.

Conway, M. A. (2009). Episodic memories. *Neuropsychologia, 47*(11), 2305–2313.

Crane, T. (2015). *The mechanical mind: A philosophical introduction to minds, machines and mental representation*. London, UK: Routledge.

Danesi, M. (2003). Metaphorical "networks" and verbal communication: A semiotic perspective of human discourse. *Σημειωτκή: Sign Systems Studies, 31*(2), 341–363.

Davies, M. (2009). The 385+ million word Corpus of Contemporary American English (1990–2008+): Design, architecture, and linguistic insights. *International Journal of Corpus Linguistics, 14*(2), 159–190.

de Saussure, F. (1972). *Course in general linguistics* (trans: Harris, R.). London, UK: Duckworth.

de Saussure, F. (2006). *Writings in general linguistics*. Oxford, UK: Oxford University Press.

Dobzhansky, T. (1973). Nothing in biology makes sense except in the light of evolution. *American Biology Teacher, 35*(3), 125–129.

Douglas, R. J., & Martin, K. A. C. (2007). Recurrent neuronal circuits in the neocortex. *Current Biology, 17*(13), R496–R500.

Eco, U. (2000). *Kant and the platypus: Essays on language and cognition*. London, UK: Vintage.

Ehresmann, A. C., & Vanbremeersch, J.-P. (2007). *Memory evolutive systems*. New York, NY: Elsevier.

Ellerman, D. P. (1988). Category theory and concrete universals. *Erkenntnis, 28*, 409–429.

Fisch, L. P., Privman, E., Ramot, M., Harel, M., Nir, Y., Kipervasser, S., … Malach, R. (2009). Neural "ignition": Enhanced activation linked to perceptual awareness in human ventral stream visual cortex. *Neuron, 64*, 562–574.

Freud, S. (1925). Negation. *International Journal of Psychoanalysis, 6*, 367–371.

Gebuis, T., Kadosh, R. C., & Gevers, W. (2016). Sensory-integration system rather than approximate number system underlies numerosity processing: A critical review. *Acta Psychologica, 171*, 17–35.

George, D. (2008). *How the brain might work: A hierarchical and temporal model for learning and recognition*. (Unpublished doctoral dissertation). Stanford, CA: Stanford University.

George, D., & Hawkins, J. (2009). Towards a mathematical theory of cortical micro-circuits. *PLoS Computational Biology, 5*(10), e1000532.

Goldblatt, R. (1979). *Topoi: The categorical analysis of logic*. Amsterdam, Netherlands: North Holland Publishing.

Golubitsky, M., & Stewart, I. (2006). Nonlinear dynamics of networks: The groupoid formalism. *Bulletin of the American Mathematical Society, 43*(3), 305–364.

Goodfellow, I., Bengio, Y., & Curville, A. (2016). *Deep learning*. Cambridge, MA: MIT Press.

Gray, E. (1995). *Hafiz the green sea of heaven*. White Cloud Press, USA.

Grill-Spector, K., & Malach, R. (2004). The human visual cortex. *Annual Review of Neuroscience, 27*, 649–677.

Grunwald, P. (2004). A tutorial introduction to the minimum description length principle. *arXiv*, arXiv: math/0406077.

Harmelech, T., & Malach, R. (2013). Neurocognitive biases and the patterns of spontaneous correlations in the human cortex. *Trends in Cognitive Sciences, 17*, 606–615.

Harries-Jones, P. (1995). *A recursive vision: Ecological understanding and Gregory Bateson*. Toronto, ON: Toronto University Press.

Hassabis, D., & Maguire, E. A. (2007). Deconstructing episodic memory with construction. *Trends in Cognitive Sciences, 11*(7), 299–306.

Higgins, J. P. (2005). Categories and groupoids. *Theory and Application of Categories, 7*, 1–195.

Historical Thesaurus of English, The. (2016). Version 4.2. Glasgow, UK: University of Glasgow. Accessed 25 Mar 2017, from http://historicalthesaurus.arts.gla.ac.uk

Hoey, M. (1991). *Patterns of lexis in text*. Oxford, UK: Oxford University Press.

Kopell, N. J., Gritton, H. J., Whittington, M. A., & Kramer, M. A. (2014). Beyond the connectome: The dynome. *Neuron, 83*(6), 1319–1328.

Lakoff, G., & Johnson, M. (2008). *Metaphors we live by*. Chicago, IL: University of Chicago Press.

Lawvere, F. W., & Schanuel, S. H. (2000). *Conceptual mathematics*. Cambridge, UK: Cambridge University Press.

Lin, H. W., & Tegmark, M. (2016). Why does deep and cheap learning work so well? *arXiv*, arXiv:1608.08225.

Luria, A. R. (1976). *Cognitive development: Its cultural and social foundations*. Cambridge, MA: Harvard University Press.

Luria, A. R., & Vygotsky, L. S. (1992). *Ape, primitive man and child: Essays in the history of behavior*. Orlando, FL: Paul M. Deutsch Press.

Lüthy, C. H., & Palmerino, C. R. (2016). Conceptual and historical reflections on chance (and related concepts). In K. Landsman & E. van Wolde (Eds.), *The challenge of chance* (pp. 9–47). New York, NY: Springer.

Macpherson, K., & Roberts, W. A. (2013). Can dogs count? *Learning and Motivation, 44*(4), 241–251.

Malach, R. (2007). The measurement problem in consciousness research. *Behavioral and Brain Sciences, 30*, 516–517.

Malach, R. (2012). Neuronal reflections and subjective awareness. In S. Edelman, T. Fekete, & N. Zach (Eds.), *Being in time: Dynamical models of phenomenal experience* (pp. 21–37). Amsterdam, Netherlands: John Benjamins.

Malach, R., Amir, Y., Harel, M., & Grinvald, A. (1993). Relationship between intrinsic connections and functional architecture revealed by optical imaging and in-vivo targeted biocytin injections in primate striate cortex. *Proceedings of the National Academy of Sciences of the United States of America, 90*, 10469–10473.

Matte-Blanco, I. (1998). *The unconscious as infinite sets: An essay in bi-logic*. London, UK: Karnac Books.

Mazur, B. (2008). When is one thing equal to some other thing? In B. Gold & R. A. Simons (Eds.), *Proof and other dilemmas: Mathematics and philosophy* (pp. 221–243). Washington, DC: Mathematical Association of America.

Mill, J. S. (1882). *A system of logic ratiocinative and inductive: Being a connected view of the principles of evidence and the methods of scientific investigation*. New York, NY: Harper & Brothers.

Moutard, C., Dehaene, S., & Malach, R. (2005). Spontaneous fluctuations and non-linear ignitions: Two dynamic faces of cortical recurrent loops. *Neuron, 88*, 194–206.

Murakami, H. (1985/1993). *Hard-boiled wonderland and the end of the world* (trans: Birnbaum, A.). London, UK: Vintage.

Neuman, Y. (2003). *Processes and boundaries of the mind: Extending the limit line*. New York, NY: Springer.

Neuman, Y. (2008). *Reviving the living: Meaning making in living systems*. Oxford, UK: Elsevier.

Neuman, Y. (2013). A novel generic conception of structure: Solving Piaget's riddle. In L. Ruddolph (Ed.), *Qualitative mathematics for the social sciences: Mathematical models for research on cultural dynamics* (pp. 255–276). Abingdon, UK: Routledge.

Neuman, Y., Cohen, Y., & Assaf, D. (2015). How do we understand the meaning of connotations? A cognitive computational model. *Semiotica, 2015*(205), 1–16.

Neuman, Y., Neuman, Y., & Cohen, Y. (2017). A novel procedure for measuring semantic synergy. *Complexity, 2017*, 5785617.

Nicod, J. (1930/1950). *Foundations of geometry and induction*. London, UK: Routledge and Kegan Paul.

Nieder, A. (2016). Representing something out of nothing: The dawning of zero. *Trends in Cognitive Sciences, 20*(11), 830–842.

Nisbett, R. E., & Masuda, T. (2003). Culture and point of view. *Proceedings of the National Academy of Sciences, 100*(19), 11163–11170.

Online Etymological Dictionary. (n.d.). Accessed 25 Mar 2017, http://www.etymonline.com

Peirce, C. S. (1902). The simplest mathematics. In *Collected papers*. (CP 4.227–323) CP 4.235.

Peirce, C. S. (1931–1966). The collected papers of C. S. Peirce. In C. Harstone, P. Weiss, & W. Burks (Eds.). Cambridge, MA: Harvard University Press.

Phillips, S., & Wilson, W. H. (2010). Categorical compositionality: A category theory explanation for the systematicity of human cognition. *PLoS Computational Biology, 20106*(7), e1000858.

Phillips, W. A., Clark, A., & Silverstein, S. M. (2015). On the functions, mechanisms, and malfunctions of intracortical contextual modulation. *Neuroscience & Biobehavioral Reviews, 52*, 1–20.

Piaget, J. (1970). *Structuralism*. New York, NY: Basic Books.

Poggio, T., & Serre, T. (2013). Models of visual cortex. *Scholarpedia, 8*(4), 3516.

Potter, D. (1990). *Hide and seek*. London, UK: Faber & Faber.

Rosen, R. (2005). *Life itself*. New York, NY: Columbia University Press.

Russell, B. (1919/1993). *Introduction to mathematical philosophy*. London, UK: Routledge.

Salomon, D. (2007). *A concise introduction to data compression*. New York, NY: Springer.

Sasso, G. (2016). The structural properties of the anagram in poetry. *Semiotica, 2016*(213), 123–164.

Schleiermacher, F. (1999). In M. Frank (Ed.), *Hermeneutik und Kritik* (7th ed.). Frankfurt am Main, Germany: Suhrkamp.

Shanon, B. (1992). Metaphor: From fixedness and selection to differentiation and creation. *Poetics Today, 13*(4), 659–685.

Smith, J. D., Flemming, T. M., Boomer, J., Beran, M. J., & Church, B. A. (2013). Fading perceptual resemblance: A path for rhesus macaques (*Macaca mulatta*) to conceptual matching? *Cognition, 129*(3), 598–614.

Spencer-Brown, G. (1994). *Laws of form*. Portland, OR: Cognizer.

Sporns, O., & Betzel, R. F. (2016). Modular brain networks. *Annual Review of Psychology, 67*, 613–640.

Stevens, M. (2016). *Cheats and deceits: How animals and plants exploit and mislead*. Oxford, UK: Oxford University Press.

Tamir, B., & Neuman, Y. (2015). The physics of categorization. *Complexity, 21*(S1), 269–274.

Terwijn, S. A. (2016). The mathematical foundations of randomness. In K. Landsman & E. van Wolde (Eds.), *The challenge of chance* (pp. 49–66). New York, NY: Springer.

Tetzlaff, C., Kolodziejski, C., Markelic, I., & Wörgötter, F. (2012). Time scales of memory, learning, and plasticity. *Biological Cybernetics, 106*(11–12), 715–726.

Timme, N., Alford, W., Flecker, B., & Beggs, J. M. (2014). Synergy, redundancy, and multivariate information measures: An experimentalist's perspective. *Journal of Computational Neuroscience, 36*(2), 119–140.

Tulving, E. (2002). Episodic memory: From mind to brain. *Annual Review of Psychology, 53*(1), 1–25.

Volosinov, V. N. (1986). *Marxism and the philosophy of language*. Cambridge, MA: Harvard University Press.

von Foerster, H. (2003). *Understanding understanding: Essays on cybernetics and cognition*. New York, NY: Springer.

Von Neumann, J. (1956). Probabilistic logics and the synthesis of reliable organisms from unreliable components. *Automata Studies, 34*, 43–98.

Wasserman, E. A. (2016). Thinking abstractly like a duck(ling). *Science, 353*(6296), 222–223.

Williams, P. L., & Beer R. D. (2010). Nonnegative decomposition of multivariate information. *arXiv*, arXiv:1004.2515.

About the Author

Yair Neuman is a Professor at The Department of Brain and Cognitive Sciences and the Zlotowski Center for Neuroscience at Ben-Gurion University. He holds a BA in Psychology (Major) and Philosophy (Minor) and a Ph.D. in Cognition (Hebrew University, 1999), and his expertise is in studying complex cognitive, social, and symbolic systems from a unique interdisciplinary approach. Professor Neuman has published numerous papers and five academic books and has been a visiting scholar or professor at MIT, the University of Toronto, the University of Oxford, and the Weizmann Institute of Science. Beyond his purely academic work, he has developed state-of-the-art algorithms for social and cognitive computing, such as those he developed for the IARPA metaphor project (ADAMA group).

© Springer International Publishing AG 2017
Y. Neuman, *Mathematical Structures of Natural Intelligence*, Mathematics in Mind,
https://doi.org/10.1007/978-3-319-68246-4

Author Index

A
Adamek, J., 31, 35
Alexander, C., 107, 109, 110, 113, 114,
    152, 160

B
Bateson, G., 11, 72, 91
Bohm, D., 74

C
Cohen, I., 15, 17, 27, 95, 114, 140, 156

D
Danesi, M, 103, 114, 158

E
Ehresmann, A.C., 31

F
Freud, S., 25, 86, 87, 89, 123, 147–149

G
Goldblatt, R., 31, 36, 37, 39, 41, 49, 80,
    81, 83
Golubitsky, M, 50, 56

H
Hoey, M., 110

L
Lakoff, G., 140, 141
Landauer, R., 25, 26, 56
Lawvere, F.W., 31, 40, 47, 49, 105, 111, 112
Luria, A.R., 71, 127

M
Malach, R., 53–55, 57, 58, 61, 62
Matte-Blanco, I., 149

N
Neuman, Y., 11, 17, 28, 35, 49, 55, 103, 114,
    140, 158

P
Peirce, C.S., 69, 88, 93, 115, 137, 138, 143
Piaget, J., 4, 7, 8, 13–29, 32, 36, 47, 50, 56,
    65, 70, 76, 91, 103, 135, 157

R
Rosen, R., 31
Russell, B., 121, 122

S
Saussure, F. de., 7, 37, 38, 75
Shanon, B., 140

T
Tulving, E., 141

© Springer International Publishing AG 2017
Y. Neuman, *Mathematical Structures of Natural Intelligence*, Mathematics in Mind,
https://doi.org/10.1007/978-3-319-68246-4

Subject Index

A
Abductive reasoning, 18, 68, 115
Adjoint, 83, 84, 113
Analogy, 4, 9, 36, 75, 90, 131, 132, 135, 136, 140, 156
Artificial neural networks, 5, 8, 10
Association, 3, 6, 66–68, 91, 104, 114, 138, 143, 153

B
Best thing of its type, 40
Boundary, 15, 20–22, 86, 91, 94, 105, 106, 114, 148, 149, 152, 158

C
Category, 11, 29, 31, 47–49, 51, 56, 72–76, 80–83, 85–87, 89–91, 101, 107, 110–115, 123, 124, 128–130, 153
Category theory, 11, 29, 31–47, 51, 56, 86, 87, 91, 111, 123, 153, 159
Center, 106–108, 110–114, 117, 160
Chance, 95–98, 159
Characteristic function, 73, 75
Closure, 20–22
Clustering, 90, 105
Combinatorial space, 18, 19, 59, 100
Composite map, 34
Context vector, 42–44
Co-product, 41–49, 55, 134–136, 146

D
Deep learning, 8, 53, 57, 66, 156
Definition, intensional, 22, 135
Difference of similarities, 74, 90
Difference that makes a difference, 72, 73, 76, 85, 89, 91, 127–129, 136
Duality, 41, 47

E
Empty set, 48, 74, 85, 122–124, 128, 129
Endomap, 111, 113, 152
Episodic memory, 115, 141, 143, 144, 146
Equivalence, 4, 29, 34, 39, 47, 49, 51, 77, 85, 88, 90, 91, 101, 104, 105, 153, 157
Equivalence of categories, 78, 90, 104
Equivalence relation, 39, 49, 88

F
Fixed point, 110–114, 117, 152, 160
Functor, 32, 34, 69, 70, 79, 80, 82–84, 100, 107, 112, 137, 157

G
Group theory, 21, 24, 29

H
Hermeneutic circle, 103, 115, 117, 160
Hierarchy, 53, 54
Hypostatic abstraction, 138, 139, 143, 146

© Springer International Publishing AG 2017
Y. Neuman, *Mathematical Structures of Natural Intelligence*, Mathematics in Mind,
https://doi.org/10.1007/978-3-319-68246-4

I

Idempotent, 111, 112
Identity, 14–16, 22, 24, 25, 28, 29, 32–34,
    38, 47, 48, 51, 66, 68–70, 77–84,
    152, 157
Identity map, 32, 33, 47, 79
Incongruence, 98, 99, 101, 102, 137, 159
Information decomposition, 59, 60
Information theory, 55, 59, 62, 89, 97, 99
Initial object, 36, 37, 46–49, 51, 74, 85,
    123, 124
Inverse function, 25, 28, 56, 62
Isomorphism, 33, 34, 37, 39, 41, 45, 46,
    48–50, 56, 66, 67, 69, 81–84, 87, 88,
    90, 91, 101, 109, 113, 114, 122,
    153, 157

K

Kullback–Leibler divergence measure,
    104–105

L

Landauer's principle, 56
Law of good Gestalt, 105
Law of identity, 24, 25, 148
Law of proximity, 103, 105
Law of similarity, 103, 105
Law of symmetry, 107
Learning by association, 6
Local symmetries, 20, 65, 107, 109, 110,
    113–115, 117, 157, 158, 160

M

Machine learning, 53, 66, 94, 100
Map, 14, 31–34, 37, 39–41, 43, 45, 47, 48, 69,
    70, 73, 75, 79, 80, 84, 87, 88, 91, 93,
    97, 99–102, 110–113, 121, 123–125,
    137, 140, 145, 157, 159
Meaning, 8, 16–18, 20, 26–28, 33, 36–39,
    42–45, 54, 58, 59, 62, 65, 67, 71–76,
    82, 85, 86, 88, 96, 98, 101, 103, 104,
    107, 115, 116, 121, 123, 128–130, 137,
    138, 140–142, 144–146, 148, 153, 157,
    158, 160
Metaphor, 5, 10, 55, 58, 59, 90, 101, 131–146,
    149, 154–156, 158, 160, 161
Modeling, 8, 9, 11–13, 18, 28, 29, 31, 32, 35,
    49, 50, 56, 61, 62, 65, 93–102, 156,
    159, 160
Morphism, 31, 32, 34, 35, 42, 46, 48, 49, 56,
    71, 72, 74, 76, 78, 87, 88, 98, 107, 113,
    115, 123–125, 129, 130, 152

N

Natural number object, 125
Natural transformation, 79–82, 84, 98–100,
    102, 107
Natural transformation modeling (NTM)
    principle, 100–102, 107, 159
Negation, 25, 85–91, 114, 123–125,
    129, 130
Neo-Darwinism, 6
Network modules, 54, 55
Number, 14, 18, 21, 44, 57–60, 75, 103, 112,
    116, 121–130, 133, 151
Numerosity, 126, 127

P

Partial function, 34
Physics of computation, 25–28, 56
Principle of symmetry, 149
Probability theory, 96–99, 102, 159
Process of generalization, 150
Product, 39–42, 44–49, 51, 55, 106, 116, 136,
    146, 156

R

Reality principle, 101
Recursive–hierarchical structure, 160
Redundancy, 55, 60, 61
Reflective abstraction, 22, 23, 35, 70,
    76, 157
Reinforcement learning, 74, 158
Relational epistemology, 31, 73
Repetition, 109, 110, 113, 151, 152
Reversibility–irreversibility, 65

S

Scales of analysis, 78, 84, 105, 108–110,
    113, 117
Second law of thermodynamics, 26
Self, 15, 16, 20–24, 29, 78, 80, 95, 104,
    124, 125, 127, 142, 146, 148,
    153, 157
Self-differentiation, 125
Self-regulation, 20, 22, 157
Semantic memory, 27, 28, 132, 138, 141,
    144, 146
Semantic transparency, 44, 45
Similarity of differences, 74–76, 90, 115, 129,
    140, 146
Simplicial complex, 88
Structuralism, 7–10, 13, 14, 65, 91
Structure-preserving transformation, 14–16,
    20, 23, 68, 69, 153, 157

Sub-object, 49, 87, 88, 103–108, 114, 143, 144, 151
Symmetric relation, 146, 148–150
Symmetry, 19, 20, 50, 106–109, 114, 149, 157
Synchronization, 50, 56, 60, 61, 91, 109, 157
Synergy, 18, 60, 114, 116

T
Terminal object, 37, 38, 47, 49, 51, 85–87, 91, 123–125
Transformation, 13–18, 20–24, 29, 31–33, 35, 67–70, 76, 77, 81, 82, 84, 107–111, 113, 116, 157, 159, 160

U
Unconscious, 42, 94, 99, 146–154

W
Wholeness, 16–18, 20, 22, 23, 29, 39, 42, 103, 107–109, 113, 116, 152, 157, 159, 160
Word compound, 18, 44, 45, 114
Working memory, 68, 151

Z
Zero, 48, 74, 85, 121, 123, 124, 127–130, 154

Printed in the United States
By Bookmasters